Ship Mo...

FROM SCRATCH

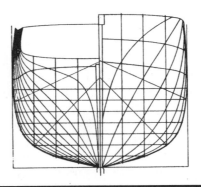

Tips and Techniques for Building Without Kits

Edwin B. Leaf

International Marine
Camden, Maine

DEDICATION

To my wife—who thinks that turning the house
into a shipmodel museum might be a good idea.

Published by International Marine

10   9   8

Copyright © 1994 International Marine, a
division of The McGraw-Hill Companies.

**Library of Congress Cataloging-in-
Publication Data**
Leaf, Edwin B.
    Ship modeling from scratch: tips and
techniques for building without kits / Edwin
B. Leaf.
        p.   cm.
    Includes bibliographical references and
index.
    ISBN 0-87742-389-X (alk. paper)
    1. Ship models.   I. Title.
VM298.L36 1993; aa21 08-30-93
623.8'201—dc20   93-35393
                                          CIP

Questions regarding the content of this book
should be addressed to:

International Marine
P.O. Box 220
Camden, ME 04843

Questions regarding the ordering of this book
should be addressed to:
The McGraw-Hill Companies
Customer Service Department
P.O. Box 547
Blacklick, OH 43004
Retail customers: 1-800-262-4729
Bookstores: 1-800-722-4726
This book is printed on recycled paper
containng a minimum of 50% total recycled
fiber with 10% postconsumer de-inked fiber.

Printed by R. R. Donnelley
Design by Ken Gross
Typeset by Farrar Associates
Production by Janet Robbins
Edited by J.R. Babb, Kathy Newman, Tom
McCarthy

CONTENTS

ACKNOWLEDGMENTS

Many thanks to Ned Brownlee, Bill Crothers, George Dukes, and Bill Wester for their invaluable assistance and encouragement. Thanks also to the members of the Philadelphia Ship Model Society whose questions and comments inspired this book. My wife, Pat, herself a novice scratch-builder, critiqued the manuscript from a novice's standpoint and took most of the photographs. Without her support, I would have never had time to write. My son Jeff, an accomplished draftsman, helped a great deal in preparing the drawings.

INTRODUCTION

Forty-six years ago I scratchbuilt a ship model for the first time—at fifteen, I couldn't afford a kit. The results weren't bad; I had an excellent set of drawings, good instructions, and a relatively uncomplicated Baltimore clipper to build. Still, it was sink or swim, and I don't recommend it.

Building a model from scratch is a singular pursuit. Selecting a project, researching its history, interpreting its lines, and choosing materials and a proper scale require patience, confidence, and ingenuity.

Unlike me, most first-time scratchbuilders have built at least one model from a kit, which is an excellent way to develop your skills. You don't have to worry about choosing materials, and the kit's drawings and instructions usually provide enough information to finish the model.

The time comes, however, when a kit-builder grows and looks for something more challenging. Scratchbuilding is the logical next step. Now *you* have to interpret the drawing and develop construction methods.

You might even have additional research and some drawing of your own to do. Certainly you'll make patterns; no more preprinted or die-cut parts.

But don't worry, the rewards are great and the choices unlimited. You, not the kit designer, control the project. You can build a model of anything you can find a drawing for—your skill level allowing, of course.

This book will ease the transition from kits to scratchbuilding, leading you from choosing a project through displaying the finished model properly. Throughout, I have emphasized the principles on which things are done rather than showing in detail how to build everything imaginable. I hope this approach helps you reach new heights in the great art of ship modeling.

Edgewater Park, New Jersey
July 15, 1993

CHAPTER 1

Selecting a Project

"We couldn't decide whether to build a four-masted clipper or a 120-gun ship-of-the-line, so we compromised on a scow."

— Ebenezer McKlutz, notorious shipbuilder

Deciding to attempt your first scratchbuilt ship is both thrilling and daunting. If you've reached this point, you've already experienced some of the hobby's intense satisfaction. For this satisfaction to continue, you'll have to choose your first scratchbuilt project wisely. You'll have to size up your skill and experience, look into the availability of information on the ship you choose, and choose an appropriate scale.

SIZING UP YOUR SKILL AND EXPERIENCE

There's much more to settling on your first project than simply picking a ship that interests you. What is your experience? What kinds of kits have you built? What are your strengths and weaknesses? If you've done only precarved hulls, you're not ready for a plank-on-frame model; try at least a plank-on-bulkhead kit first, or perhaps a small-boat kit that requires planking. Do you have trouble with elaborate rigging? Don't attempt a project that will frustrate you. Too many beginners have attempted the *Flying Cloud* or the *Constitution* and have become so discouraged they've sworn off ship modeling forever. Try a smaller ship or some type of small craft — many are very satisfying, even for an expert. If you build the *Constitution*, it will be only one of thousands. If you choose a small, unusual subject, your model might be unique. Try a simple schooner, or look into vessels significant to your area. Consider also the boredom factor. If you choose too ambitious a project, you might tire before it's finished.

Figure 1-1, for example, is a block sloop from the War of 1812, a simple-looking model of an historic and unusual craft. It's an excellent project — and an excellent example of a common trap: It's much more difficult to build than it looks. In fact, only an experienced modeler should attempt the quite difficult construction of the unusual hull; it's not easy to cope with this sloop's bluff ends, abrupt curves, and the integration of the turtlebacks at the bow and stern.

Consider also how much time and aggravation you're prepared to invest, for you'll have to live with your decision for a long time. Even small vessels can be very challenging. Take, for example, the block sloop. I studied the drawings, analyzed them, and decided that a month or so would do the job. More than three months later, after overcoming loads of unforeseen complications, the model was finally finished.

Figure 1-2 is a New England pinky schooner, a common fishing vessel used in the nineteenth century, and a uniquely

Figure 1-1. This unusual model of an American block sloop from the War of 1812 is full of pitfalls for the beginner. (Photo by Patricia Leaf)

American ship with an interesting history. Not many models are built of such little ships, and it would be a good project for a beginner with some woodworking skills. At ⅜″=1′ scale, its detail is satisfying yet not too painstaking, and the plank-on-frame hull is easy to build because of its gentle curves and sweeping sheer.

Figure 1-3 shows a Block Island boat, or "cowhorn." It's a classic American craft, not much modeled, and an excellent first lapstrake project.

Figure 1-2. The pinky schooner is an historically important American type, and an excellent first project. (Photo by Patricia Leaf)

Figure 1-3. The Block Island boat, or cowhorn, is a distinctive American type, and a good choice for a first lapstrake project.

AVAILABILITY OF INFORMATION

What information is available on your model of choice? Although kits usually contain all the information you need, scratchbuilding means that you might have to do your own, often extensive, research. If you're not ready to invest hours at the library, choose a project for which detailed and authoritative drawings are available. Don't attempt to build a model with inadequate information. For example, little is known about Columbus's *Santa Maria* or the *Mayflower*, or about a host of other famous ships. In some cases, there are hypothetical reconstruction drawings, or even full-scale "replicas" of such ships. If you want to build a model of one of these, and you have the data, go ahead, but be sure you label your model for what it is — an informed conjecture.

The model in Figure 1-4, for example, purports to be a Phoenician galley. It probably should not have been built, except as a curiosity. Careful study of the model shows that such a heavy ship could never have been propelled by oars. The model's plans

Figure 1-4. This model purports to be of a Phoenician galley, but the plans were based on misinterpreted information. (Photo by Gregory Masters)

were based on an ancient relief carving that was seriously misinterpreted.

Evaluate all plans carefully, no matter the source; some plans might be difficult to interpret or lack essential details. Avoid drawings that confuse you. Lack of detail is a problem common to drawings dating from the Age of Sail, which frequently omitted deck arrangements and rigging details that often were done according to rules of thumb and conventions no one thought necessary to record. Working from such drawings requires research. Written descriptions might be available, and from the mid-nineteenth century on, photos might exist. Look at more complete drawings of other similar ships of the same period. Contemporary artists' renderings may or may not be useful. Many were done by artists who were not knowledgeable about ships, or who drew according to a contemporary style that was not wholly realistic. Nineteenth-century prints of the Currier and Ives type typify this fault, as do seventeenth- and eighteenth-century popular prints. Twentieth-century standards, however, are high, and older drawings (such as those of Van de Velde) are superb. Appendix II lists books that tell you how things were done at different times and places. Use everything you can find, and be careful not to incorporate any inappropriate or anachronistic detail.

Read and study as much as possible; the best sources are those that talk about real ships and seamanship, not about models. Seek out primary sources — you need to understand your ship and how it was really built. The best drawings are those actually used in the construction of the ship. Drawings made especially for modeling are often more confusing than not, since they tend to condense or even omit information. Maritime museums, of which there are many throughout the country, usually maintain useful libraries and have exchange arrangements with other museums. If there is such an institution available to you, take advantage of it. Chapter 2 discusses in detail how to use drawings.

A final word: If you're not sure about a detail on the ship, ask yourself if it would have worked. Ships were, and are, extremely functional. If something looks impractical, it's probably wrong.

Your primary concern should be developing your skills. Don't be too concerned about absolute accuracy. If you're working from reasonably reliable plans, go ahead and don't worry about extensive scholarly research. Rome wasn't built in a day, and you're not expected to produce a masterpiece on your first try.

CHOOSING AN APPROPRIATE SCALE

First, a word or two about the meaning of scale. Scale is the ratio of the size of the model to the size of the real ship. Scale can be expressed in two ways, both of which you will encounter. The actual ratio of the model to the real ship might be used — for example, 1:48 means the model is $\frac{1}{48}$ the size of the original. Since this is not really convenient for the model builder, we prefer to work with an expression that converts the ratio to inches and feet — for example, $\frac{1}{4}''=1'$ tells us that $\frac{1}{4}$ inch on the model is equal to 1 foot on the actual ship. The scale 1:48 is the same as $\frac{1}{4}''=1'$.

Choosing the scale for a particular model involves both practical and aesthetic considerations. Since the scale determines the size of the model, you must first take into consideration the space available both in your workshop and in the intended display area.

The following table shows the finished sizes of various ships at different scales.

Note the large size of the battlecruiser at a scale of $\frac{1}{8}''=1'$, or of the ship-of-the-line at $\frac{1}{4}''=1'$. Few people have the space, either in the workshop or in the display area, to indulge in such gigantism, so the smaller scale is preferable. On the other hand, the catboat at a sale of $\frac{1}{4}''=1'$ would not be a very impressive model, unless you like miniatures. A larger scale, even as much as $1''=1'$, would be preferable for the catboat.

Another consideration is the level of detail desired. Obviously, the larger the scale, the easier it is to incorporate a great deal of detail. If you don't want to work on very tiny pieces, it would be a good idea to build models of small vessels at a large scale. In the long run, the decision is up to you. Whatever you choose, you might have to adjust your drawings to the desired scale (covered in Chapter 2).

Once you select the scale, you need to adjust the level of detail on the model to the scale. At $\frac{1}{16}''=1'$, a hatch reveals little detail; but at $\frac{1}{4}''=1'$, almost every detail needs to be shown. Include as much detail as your skill allows without cluttering the model. This means that everything must be kept in scale. An eyebolt with a model diameter of $\frac{1}{16}$ inch would be 6 inches in diameter on the real ship at $\frac{1}{8}''=1'$, and this is ridiculous. Practical considerations may make it difficult to do much about this particular item;

Ship	Actual Size	Scale	Ratio	Model Size
Battlecruiser	650'	$\frac{1}{8}''=1'$	1:96	81.25"
Battlecruiser	650'	$\frac{1}{16}''=1'$	1:192	40.625"
Ship-of-the-line	275'	$\frac{1}{4}''=1'$	1:48	68.75"
Ship-of-the-line	275'	$\frac{1}{8}''=1'$	1:96	34.375"
Catboat	30'	$\frac{1}{2}''=1'$	1:24	15"
Catboat	30'	$\frac{1}{4}''=1'$	1:48	7.5"

you do have to use the eyebolt to terminate a line. A doorknob on a cabin door would have to be less than $\frac{1}{32}$ inch at $\frac{1}{8}''=1'$. Is it worth including? Looking at a model is a bit like looking at a real ship from a distance. Would you really be able to notice that doorknob? Probably not. If you would need a magnifying glass to appreciate a bit of detail, it might be better to leave it off.

Some modelers choose to build a series of models, such as of classic boats from a particular area or time. In this case, it's best to choose one scale for all models in the series to show relative size.

REFERENCE MATERIALS

There are a great many books on the market that purport to tell you how to build models. Some are valuable, others are not. Some authors describe techniques that they must never have actually tried, because they just won't work. Read "how-to" books (including this one) with a critical eye. There are many valid ways of doing things. Decide what works for you, and forget the rest.

Appendix II lists selected references with a brief description of each. These books are excellent resources, but some are, to a degree, outdated. For example, the glues and finishes available to us today are superior to those used a generation ago, and this can change the way we do certain things.

Standards have changed, too. Some things recommended fifty years ago are not acceptable today. Scribing decks instead of laying separate planks, cutting out the whelps of a capstan with a penknife, and fastening copper plates with small nails are examples of what are considered today unacceptable practices. Be suspicious of any procedure that "represents" rather than replicates the original.

THE MYTH OF "MUSEUM QUALITY"

Much is made of the term "museum quality," which somehow seems to imply the epitome of high standards for a model. The U.S. Navy and Mystic Seaport both maintain modeling standards addressing the use of materials and methods of construction; they do not, however, address the question of whether the model is an accurate representation of the ship. Go to just about any maritime museum and you'll see many models that are not of exceptional quality; they were chosen instead for other reasons, such as historical value or as a specimen of an unusual vessel. If you're concerned about the quality of your work, choose materials that are appropriate and enduring and do the best you can. There is no better standard for the hobbyist.

CHAPTER 2

Getting the Most from Your Drawings

"I just can't understand why you find these drawings hard to read. I used a great new system of drawing the lines — my very own invention!"

— Peter Pen, undistinguished naval architect

If you've already built a ship from a kit, you have some experience reading and interpreting drawings. Kit drawings, however, are often simplified or redrawn to reflect a particular type of construction. With scratch-building, you deal with more sophisticated drawings, including reproductions of the original drawings from which the real ship was built. You interpret them and adapt them to the type of construction you choose. This chapter reviews the information contained on a drawing and discusses the techniques for changing scale and enhancing the drawings.

INFORMATION CONTAINED ON A DRAWING

The shape of a ship's hull is represented on a drawing in three views: plan (or half-breadth), elevation (or profile), and body (or section). Each view shows, from different perspectives, waterlines, buttock lines, diagonals, and sections, all of which define, in different ways, the form of the hull. Figure 2-1 shows a typical drawing, annotated to point out these lines. On builder's drafts

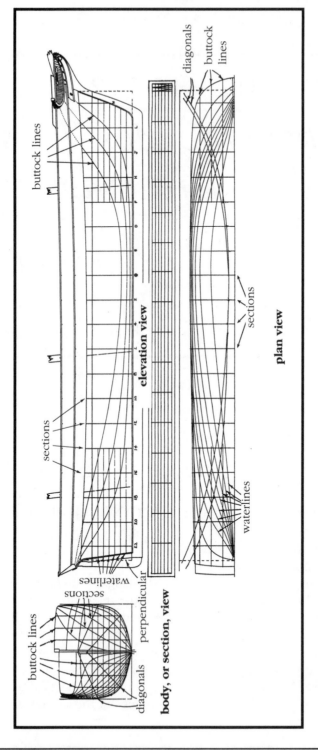

FIGURE 2-1. You might work from a drawing like this, a typical builder's draft. (Based on a drawing from *The History of American Sailing Ships*, W. W. Norton & Co.)

these lines are usually drawn to the inside of the planking (or plating). Modern naval architects, however, usually draw to the outside of the planking. Many model drawings are also drawn to the outside of the planking. Look on the drawing to determine how your plan is drawn.

The *plan view*, usually depicting only half the hull (half-breadth), shows the shape of the waterlines and diagonals, with the buttocks and sections shown as straight lines. The *elevation view* shows the shape of the buttocks, with the waterlines and sections shown as straight lines. The *body plan* shows the shape of the sections, with the buttocks, waterlines, and diagonals shown as straight lines. Usually, the right half of the body plan shows half-sections from the midsection to the bow, and the left half shows half-sections from the midsection to the stern. The midsection is often marked with the symbol " ⊗," with forward sections lettered and after sections numbered for identification. You'll find variations on this system, especially where the sections are numbered according to their distance from

the bow — for example, section 325 would be 325 feet from the bow reference line.

WATERLINES

Waterlines represent horizontal planes passed through the hull (Figure 2-2). If you're building a lift model, the waterlines are of great importance since they define the shapes of the lifts (the layers of wood used to build up the hull). You might have to interpolate to obtain the shapes of the lifts you need (see "Enhancing the Drawing," later in this chapter). The waterlines are not always shown parallel to the keel or to the load waterline. That's why I build certain lift models using the buttock lines rather than the more conventional waterlines.

BUTTOCK LINES

Buttock lines represent vertical planes passed through the hull parallel to the centerline (Figure 2-3). These lines are rarely used in the process of constructing a model, but are useful in checking the shapes of the

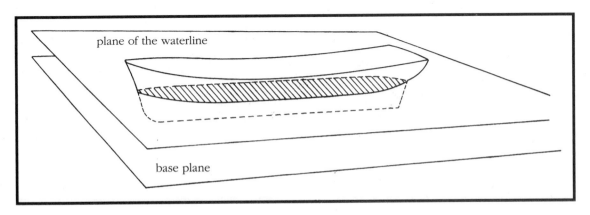

FIGURE 2-2. Plane of a waterline.

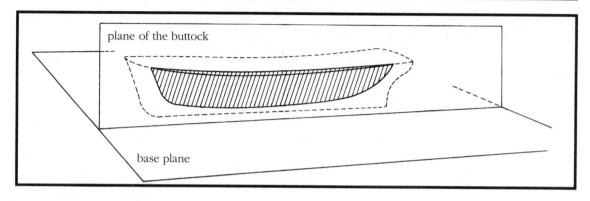

FIGURE 2-3. Plane of a buttock line.

sections. Their true shapes are shown on the elevation view. On the section view, they are vertical straight lines; on the plan view, they are horizontal straight lines.

DIAGONALS

Diagonals represent planes passed through the hull at angles to the base plane and there-fore to the vertical plane of the centerline (Figure 2-4). Although they serve no purpose in actual construction, diagonals are invaluable in checking the accuracy of the other lines (see "Enhancing the Drawing," later in this chapter). On the section view, they are shown as straight lines, and their shapes are usually shown as an overlay to the plan view (refer again to Figure 2-1).

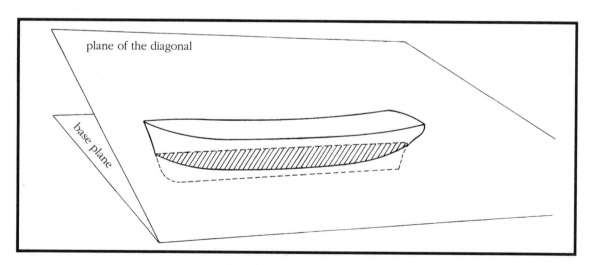

FIGURE 2-4. Plane of a diagonal.

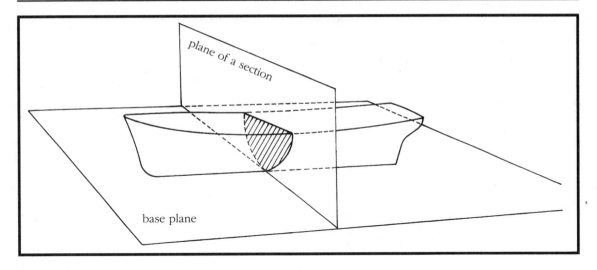

FIGURE 2-5. Plane of a section.

SECTIONS

Sections represent vertical planes passed through the hull perpendicular to the centerline (Figure 2-5). On the elevation and plan views their locations are usually referred to as stations. The sections are especially important since they represent the shapes of frames and define the shape of the hull more graphically than do the buttocks and the waterlines. Rarely, however, does the drawing show the shape of every frame, and the sections as drawn might not be at the actual locations of the frames as required or as built. Except in the case of some warships where frame locations can be rather odd in order to get around gun ports, this shouldn't concern the modeler.

On the drawing you might see a section (station) identification scheme similar to this:

19 17 15 12 9 6⍉C F I L N P

The numbers and letters represent sections that are shown; missing numbers and letters represent sections that are *not* shown. The sections shown on a drawing are rarely enough to build a plank-on-frame model, so you'll have to interpolate to determine the shapes of additional sections (covered in "Enhancing the Drawing," later in this chapter). You should, however, be able to make templates for a lift model from the sections provided.

The sections on the drawing indicate a plane, but the frames have thickness. For models, it's safe to assume that, except for the midsection, the sections represent the midsection side of the frames. This enables you to cut a frame to the shape of the section and then bevel it toward the bow or stern as required.

OTHER LINES ON THE DRAWING

Other important lines on the drawing include reference lines, rabbet lines, and the bearding line.

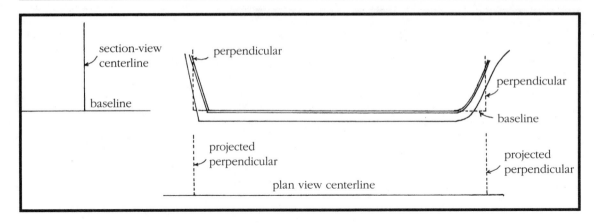

FIGURE 2-6. The drawing is built around the reference lines.

The primary reference lines are the baseline and perpendiculars of the elevation view (Figure 2-6) and the centerlines of the plan and section views. The baseline is extended to the section view, and the perpendiculars to the plan view.

The baseline appears on the elevation and section views. It is not always parallel to the keel, especially when the keel has "drag" toward the stern. The perpendiculars are often placed at the fore- and aftmost points of the stem and stern rabbets, though the forward perpendicular might be found at the intersection of the load waterline and the stem rabbet. Although the placement at times seems arbitrary, there is always a rationale for it.

The perpendiculars appear on the elevation and plan views. They are perpendicular to the baseline, which means they are not necessarily perpendicular to the keel. The length of a ship is frequently expressed as the length "between perpendiculars" (also known as "carpenter's measure"). The rabbet lines show the intersection of the planking with the keel, stem, and sternpost.

The outer rabbet line defines the intersection with the outside of the planking; the inner rabbet line defines the intersection with the inside of the planking (Figure 2-7).

Figure 2-7 also shows the cross-section of a typical rabbet, which is cut into the keel, stem, and sternpost. The width of the rabbet depends on the thickness of the plank and the angles at which the planks intersect the rabbet.

If the quality of the drawing is high, it might contain other useful information, such as the bearding line (Figure 2-8). The bearding line is an extension of the inner rabbet at the stern from which the deadwood is tapered toward the aft end of the keel rabbet and the sternpost rabbet. This is necessary so that the planking approaches the outer rabbets smoothly. If the bearding line is not shown, you can derive it from the points at which the lower ends of the sections near the stern intersect the deadwood.

The drawing might also show the complete shapes of the keel, keelson, deadwood, sternpost, stem, and deck beam locations (Figure 2-9).

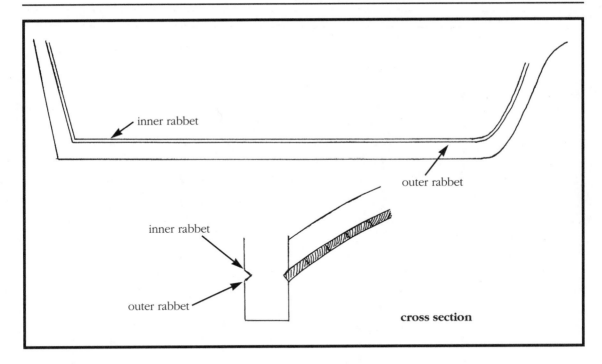

FIGURE 2-7. Rabbets.

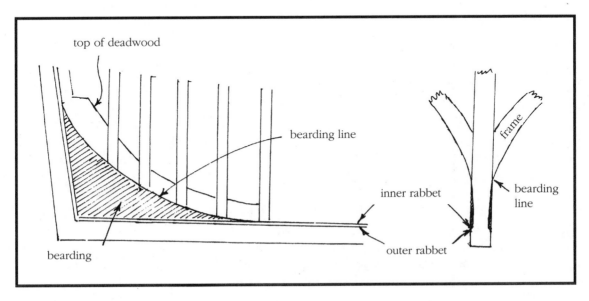

FIGURE 2-8. Bearding line.

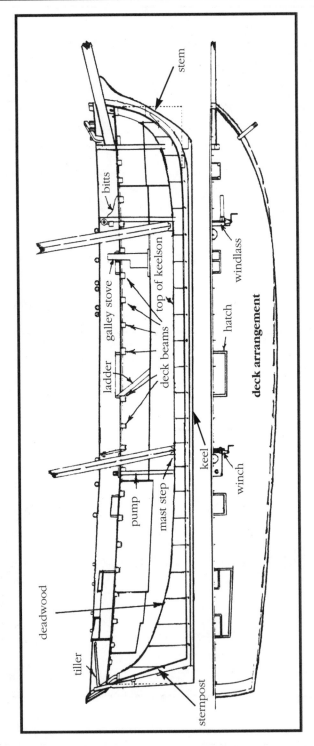

FIGURE 2-9. Your drawing might not always show as much detail as this example. (Based on a drawing from *The History of American Sailing Ships*, W. W. Norton & Co.)

OTHER USEFUL INFORMATION ON THE DRAWING

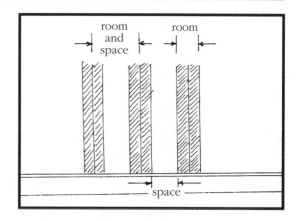

FIGURE 2-10. Room and space.

The notes included on the drawing might specify the dimensions of room (the thickness of the frames) and space (the distance between them). See Figure 2-10. Together these measurements represent the distance between the centers of the frames. This will give you a good idea of the number of frames actually used in the construction of the ship. Unless you plan to leave part of the frame unplanked, however, you don't have to use so many frames. For a fully planked model, you need only enough frames for a spacing of 1 inch or so between centers.

The notes might also include the ship's measurements, timber sizes, spar dimensions, and the source of the drawing. The scale of the drawing will always be indicated. It might be a simple notation such as $\frac{1}{4}'' = 1'$ (1:48), or it might be represented by a graphic scale, a portion of which is shown in Figure 2-11. The divisions starting at zero and moving to the left represent feet. Pay close attention to the portion at the right that shows how to measure in inches. Note that the diagonal lines intersect six intervals. These intersections represent inches.

Other information found on drawings,

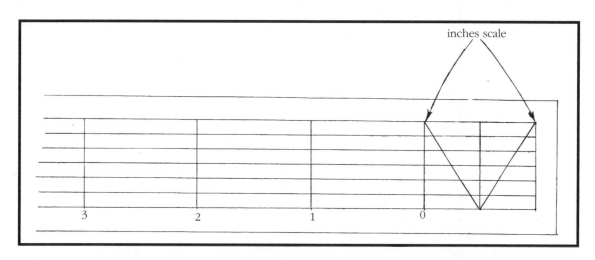

FIGURE 2-11. The diagonal lines at the end of the scale measure inches where they intersect the horizontal lines.

such as the deck arrangement and deck furnishings, is easy enough to interpret and need not be discussed here. For these, you'll primarily take measurements and make tracings for patterns.

CHANGING THE SCALE IN A DRAWING

All too frequently the available drawings are not in the scale you want, so you have to enlarge or reduce them. If you have access to a copier that can do enlargements and reductions, you can use it to do the job. However, if you redraw them yourself, in the process you will become thoroughly familiar with the ship and with any problems you might encounter in building the model. And don't be surprised if, in the process, you uncover inaccuracies in the original drawing. More about this later.

You need not modify all the details in the original drawing. In most cases, modifying the hull lines suffices (you need the hull lines to make the patterns for cutting the wood). Almost everything else can be done by taking measurements directly from the original drawing. Don't do more than you have to. It's never necessary to enlarge or reduce a rigging plan — only the sail patterns.

Start with the scale of the original drawing. If the scale is specified, you can easily double it simply by multiplying all the dimensions by 2 (for example, going from ⅛"=1' to ¼"=1'). But it isn't always that easy. A drawing in a book might have been reduced to an odd size to fit the page and you find that the scale doesn't correspond to anything even.

I've found it best to use the grid in the original drawing to change the scale, even if it's awkward. Starting with the elevation view, you would create a new grid to the desired scale as follows:

1. Establish the elevation-view baseline and perpendiculars in the desired scale.
2. Add the stations for the sections and the waterlines to the elevation view.
3. Establish a centerline for the plan view.
4. Project the stations from the elevation to the centerline of the plan view.
5. Add the buttock lines to form the plan view grid.
6. Establish the centerline for the section view.
7. Project the waterlines from the elevation drawing to the section view.
8. Add the buttock lines to the section view, taking their spacing from the plan view.
9. Add the diagonals to the section view.

You might also want to add additional waterlines, buttocks, and sections both to the original and enlarged grids for better accuracy (covered in "Enhancing the Drawing," later in this chapter). The original drawing probably won't show waterlines above the load waterline since they would confuse the drawing of the upperworks. Look closely, though, because you might find tick marks on the stations that show where they would run. Finally, add the diagonals to the section view. The result will look like Figure 2-12.

Now you're ready to plot the lines. Find points at which the lines intersect the grid on the drawing, transfer the points to the new grid, and connect the points smoothly. Be sure to include the lines of the rails, plank sheer, wales, and gunport and cathead locations.

Another way to change the scale of a

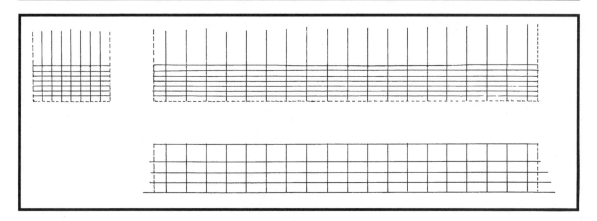

FIGURE 2-12. Use a grid like this one for enlarging or reducing drawings.

drawing is with a computer scanning program. Simply scan the image into the computer, crop the image to reflect known dimensions of the original drawing, instruct the computer to enlarge the image to the desired size, and print it on a printer-plotter for best results. This is a quick and easy method that I have used a number of times with great success. Figures 13-1 and 14-1 were developed using this technique.

CHECKING FOR ACCURACY

When you've completed your new drawing (now in the desired scale), check the lines for accuracy with diagonals. Since the diagonals reveal the accuracy of the sections, plot them directly on your drawing rather than enlarging them from the original. Smooth, evenly curved diagonals mean the frames can be planked evenly. Deviations indicate sections that are too big or too small in places. Take measurements from the centerline along the diagonals to their intersections with each section on the body plan and plot them from the baseline, overlaying the plan view (Figure 2-13).

You determine the ends of the diagonals by projecting the points P to the sternpost and stem rabbets and then dropping perpendiculars to the plan view (Figure 2-14).

Add more diagonals to your drawing for additional checks on your accuracy. For some reason, drawings of modern ships rarely show diagonals. Don't trust that a drawing is accurate; check it for your own peace of mind. When you've plotted the diagonals, go ahead and make adjustments, first to the sections and then to the waterlines as required.

Now that you've changed the scale of the drawing and checked it for accuracy, you can go on to enhance it if you wish.

ENHANCING THE DRAWING

Which elements you decide to enhance on a drawing will most likely depend on the type of construction you chose. For a plank-on-frame model, you'll probably want to

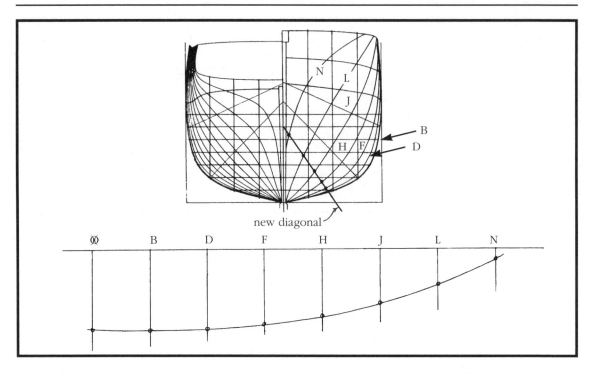

FIGURE 2-13. Plotting a diagonal.

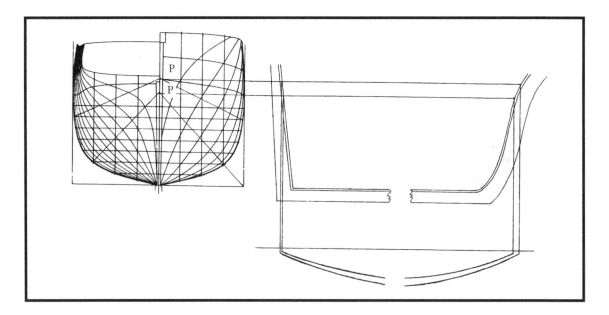

FIGURE 2-14. A method for finding the ends of diagonals.

add sections for additional frames (such as for cant frames). For a lift model, you might draw new waterlines to accommodate the thickness of the wood to be used, or additional ones for improved accuracy. For a carved or lift hull, for which the plans are drawn to the *inside* of the planking, you'll have to *pad* the sections and waterlines to compensate for the thickness of the plank. And for a plank-on-bulkhead or a plank-on-frame model, for which the plans are often drawn to the *outside* of the planking, you'll have to *reduce* the lines by the thickness of the plank.

As stated earlier, most drawings show only selected sections. Since these are not usually enough to support the planking, you'll need to determine the shapes of the others required. Furthermore, drawings almost never show the cant frames (the half-frames not at right angles to the centerline located near the bow and stern). Because of the increasing curvature at the ends, cant frames are needed if the planks are to lie at a reasonable angle to the frames. They're also needed to reduce the bevel required on the frames. Cant frames might supersede some of the foremost and aftermost sections shown on the drawing. You might also want to re-space frames to accommodate gunport locations or for other reasons.

Begin by adding new stations to the elevation and plan views. Spacing can be determined by the specified room and space, or by spacing equally between existing sections. For example, if sections B and F are shown, C and E are missing and can be spaced equally between B and F (Figures 2-15 and 2-16). Cant frames might be quite close together at the inboard edges, with the outboard edges more or less evenly spaced (Figure 2-17).

You can then plot the shape of the new frames on the body plan by taking measurements along the new lines from the centerline to the waterlines on the plan view, from the baseline to the diagonals, and (except for the cant frames) from the baseline to the buttocks on the elevation view. Refer to the bearding line to find the lower ends of new frames near the stern. If the original drawing doesn't show the bearding line, you can develop one by plotting the lower ends of the aft sections on the elevation view. Check your work using the diagonals. You cannot guess at interpolated frames by drawing "averaged" lines between existing sections.

With the additional frames plotted, work out the full shapes of the keel, stem and sternpost, deadwood, and keelson if you decide to use one (refer again to Figure 2-9 for an example of a keelson). Using a keelson is neither necessary nor common. Most modelers simply notch the keel (made deeper for the purpose) to receive the frames (Figure 2-18). More will be said about this in Chapter 6.

Enhancing the waterlines for a lift model is done in much the same way as enhancing the sections. If the waterlines are parallel to the baseline, establish the new waterlines you want on the section view. Plot them on the plan view by taking measurements from the centerline of the section view to each section in turn (see Figure 2-19). If the waterlines are not parallel to the baseline, they will appear curved on the section view. If this is the case, establish the new waterlines on the elevation views and plot them to the section view by taking the height of the waterline at each section. This is more complicated than if the waterlines were parallel to the baseline, but it isn't an uncommon situation. Determine the end points of the new

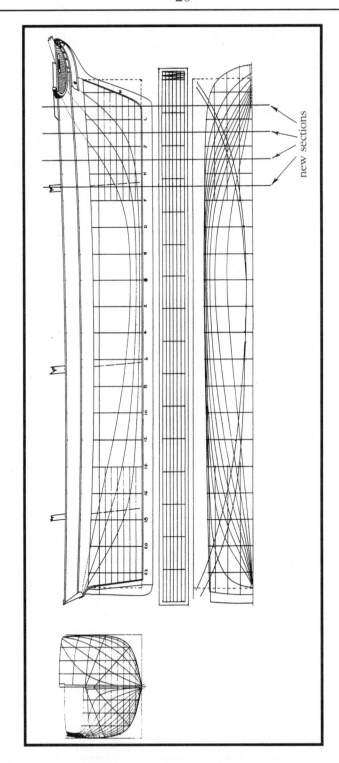

new sections

FIGURE 2-15. Locating new sections.

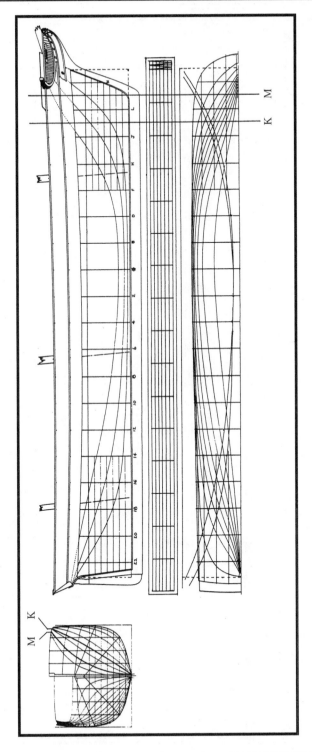

FIGURE 2-16. Plotting sections.

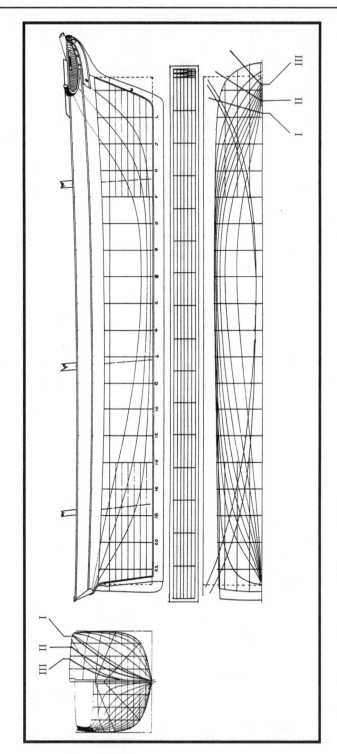

FIGURE 2-17. Developing cant frames.

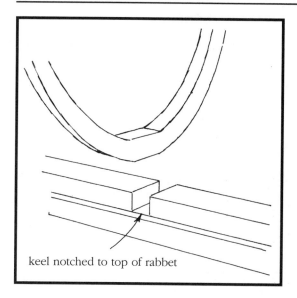

keel notched to top of rabbet

FIGURE 2-18. The keel and the frames interlock for strength.

waterlines by projecting the points at which they intersect the stem and stern rabbets on the elevation view to points half the thickness of the keel above the plan-view centerline.

By now you should have a drawing with all the information you need to start cutting wood. Use tracing paper to make patterns from the drawing (a light box could come in handy here) and transfer them to the wood. Do this by blackening the back of the paper with a soft (3B to 5B) pencil. Lay the paper on the wood with the black side down, and trace the lines with a hard pencil (3H to 4H). Don't use carbon paper — it leaves a greasy line that's hard to remove. Another way to make a pattern is to photocopy the drawing for each frame or lift required, cut it out, lay it on the wood, and draw around it.

If you decide to use proper built-up frames, you'll need to do some additional work to fully develop the inside shapes of the frames. You can also develop the outside bevels for each frame by plotting a section for the front or back of each frame, though this is quite ambitious. Just remember that forward of the midsection the sections represent the aft side of the frame (F1), and aft of the midsection, vice versa. Simply establish new sections to represent the thickness of the frames (F1–F2) (Figure 2-20).

Although this is a lot of work, it's worth the effort if you're accurate, especially for frames that have an extreme bevel or whose bevel is not constant. If you want to pre-establish the inside bevels of the frames, plot the outlines of the insides of the frames on the section view, and proceed as for the outside bevels. This is only worth doing, though, if the frames are to be exposed, as on Admiralty models, or if the insides of the frames are to be planked. Chapter 6 has more to say about shaping frames.

ENSURING TRUE SHAPES

A three-view drawing doesn't always represent the shape of certain items accurately. A sloping, curved transom is a common example. The shape of the transom on the body plan doesn't represent the true height because of the slope, nor the true width because of the curve. The actual dimensions, however, are noted on the elevation and plan views, respectively. The challenge is to create an accurate expansion by means of a grid superimposed over the body-plan view of the transom.

First, be sure that the three views on your drawing are conventionally arranged, as shown in Figure 2-1. If they're not, trace the views and then arrange them in the customary way. This is necessary so that you

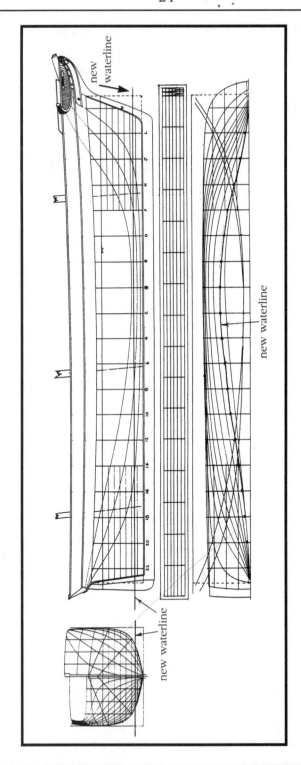

FIGURE 2-19. Developing waterlines.

can project from one view to another. It's a good idea to make a tracing, anyway, so that you don't draw your construction lines over other lines or text on the original drawing. It also avoids confusion.

Refer to Figure 2-21 throughout the following discussion. Establish a grid over the body-plan view of the transom by extending the buttock lines over it, adding vertical lines at the ends of the transom, and then drawing a number of equally spaced horizontal lines over it. These horizontal lines will be extended to intersect the elevation view of the transom.

Next, from these intersections, draw lines perpendicular to the elevation view. Extend the right-hand vertical on the body plan to intersect the topmost perpendicular to the elevation view. From this intersection (point A), draw a line at right angles to the perpendiculars from the elevation view. This line will be the right-hand side of the adjusted grid. Find the true width of the transom by measuring the curve shown on the plan view with a thin piece of wood (a spline) curved over the drawing. Measure the true width from the right side of the adjusted grid to the left, and draw another line at right angles to the perpendiculars from the elevation view. Extend this line to intersect the vertical line at the left of the body plan (point B). Draw the line AB. Now extend all the verticals on the body plan to line AB, and from the intersections draw lines to complete the adjusted grid. You can now plot the true shape of the transom on this grid.

An excellent reference for drafting and lofting in general is *Boatbuilding* by Howard I. Chapelle (see Appendix II for publishing information).

ENHANCING DRAWINGS FROM PHOTOGRAPHS

From time to time you might find that a drawing omits significant details that often show up on a good photo of the ship. Figure 2-22, for example, is a lines drawing of *Novelty*, an attractive fishing boat. This would make a good model, but the drawing is obviously incomplete. Fortunately, a photo of the ship is available (Figure 2-23). Granted, there may be some distortion of scale in the photo, but it's relatively small and you should be able to extrapolate many of the details using basic principles of perspective.

To make this easier to follow, I'll use a simplified tracing of the photograph to illustrate the method (Figure 2-24). The underlying perspective principle is that parallel lines appear to intersect at a point on the horizon (unless they are themselves parallel to the horizon, in which case they intersect at infinity). These points are called *vanishing points*, and our first step is to find them for our photo.

Find two parallel fore and aft lines on the drawing of the vessel — the top of the deckhouse and the centerline of the ship at the waterline are such lines. Extend these lines to the right until they intersect; their point of intersection will lie on the horizon, and will be the first vanishing point (labeled v.p.1 on Figure 2-24). Draw a horizontal line (the horizon) through this point. Now find an athwartships line on the ship; the front edge of the deckhouse is suitable. Extend this line to the right until it intersects the horizon line; this is the second vanishing point (v.p.2). As it happens, this second point is not used in this demonstration, but you need

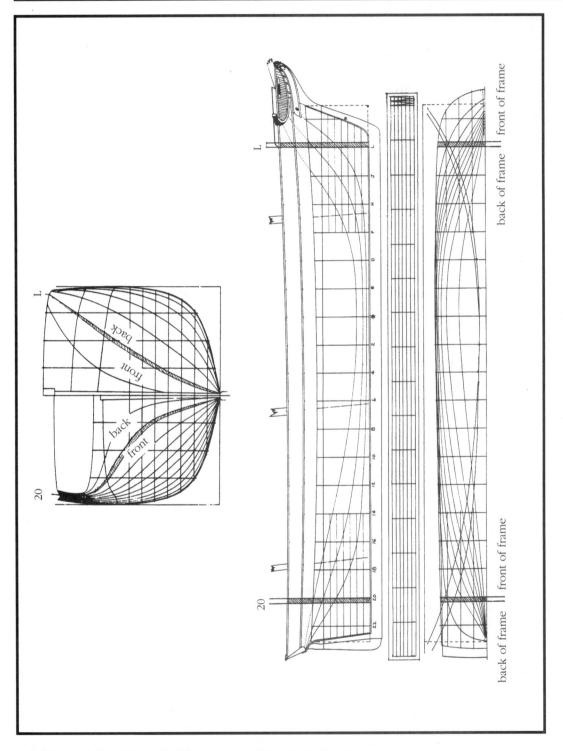

FIGURE 2-20. Plotting a bevel.

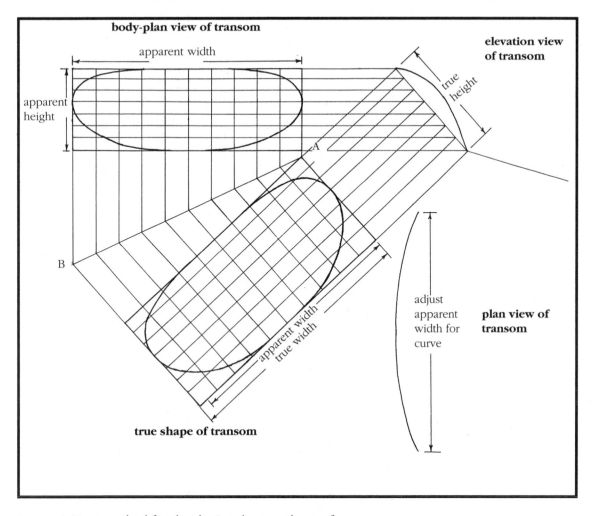

FIGURE 2-21. A method for developing the true shape of a transom.

to know how to find it because it might be needed in some of your own applications.

Next find something on the vessel whose size you can estimate or measure for the drawing. In this case, there's a man standing on the wheelhouse. We could use something else, but he's convenient. It's probably fair to say that he is about 5′ 9″

tall. Run lines from his head and feet to the right-hand vanishing point (Figure 2-25).

You can now use a principle of perspective to determine the heights of the smokestack and masts. The principle states that all vertical lines drawn between two lines leading to a vanishing point are equal in length. Therefore, a vertical line drawn

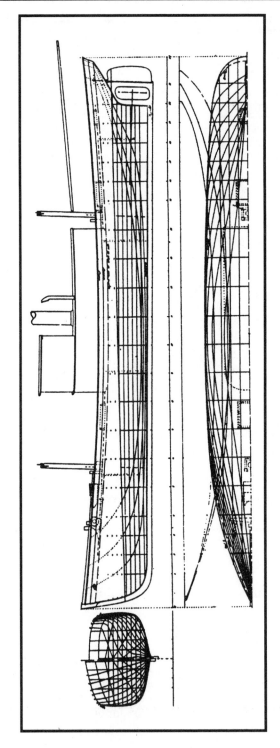

FIGURE 2-22. This lines drawing of *Novelty* lacks significant information. (Courtesy International Marine)

FIGURE 2-23. This photograph of *Novelty* can supply needed information. (Courtesy Smithsonian Institution)

between the lines in Figure 2-25 at the position of the smokestack is 5′9″ long, or one unit. The height of the smokestack from the top of the deckhouse is 4⅔ units times 5′9″, or 26′10″.

You can use this same procedure to measure the masts (Figure 2-26). Extend a pair of lines through the head and foot of the man and through the fore and main masts to the vanishing point at the right. The intersections with the masts give you the height of the man at the masts. Now you can use these units to calculate the heights of the masts.

By a somewhat more complex procedure, you can also find the length of the gaff on the foremast. Refer to Figure 2-27 for the following discussion. (The diagram at the bottom of the figure is expanded for demonstration purposes only.) Draw a line BV through the top of the gaff (B) to the right-hand vanishing point.

Remember the basic rule of perspective regarding lines that intersect at a vanishing point: BV and the horizon are parallel and form your primary reference lines. Now drop a line from the top of the gaff (B) to the horizon line, and label this intersection (E). Line BC represents the length of the gaff, which we want to find. Line AD represents the length of the portion of the foremast from the horizon line to the height of the tip of the gaff (B). Again, by the rules of perspective, lines BE and FG are equal to AD, and the angle CAB is a right angle even

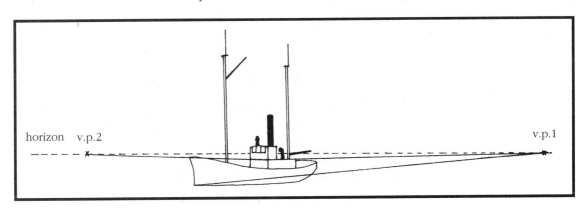

FIGURE 2-24. Locating the vanishing points.

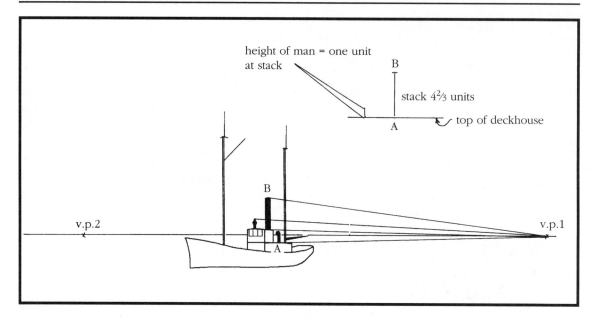

FIGURE 2-25. Finding the height of the stack.

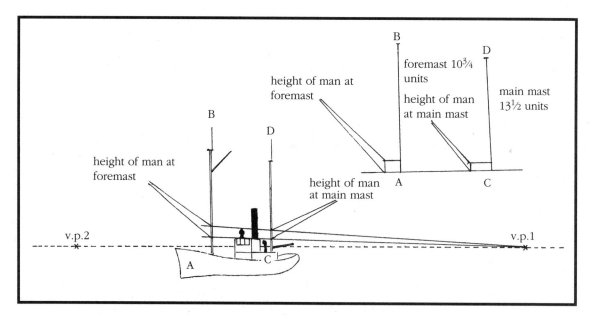

FIGURE 2-26. Finding the height of the masts.

though it does not appear so in the diagram. The points ABED therefore form a rectangle; as opposite sides of a rectangle, AB must equal DE. You can now find the length of AC in the same way you found the height of the smokestack.

Now, if you can find the length of DE (which equals AB), you can solve the triangle ABC for the length of the gaff (BC) using the Pythagorean Theorem. Find DE by approximation. The intersection of the lines AG and DF is the center of the rectangle AFGD, and a vertical line HI through this in-

tersection bisects DG at I so that DI equals ½ DG. Doing the same thing to the rectangle ADIH, we draw a vertical line KL. We now see that DL equals ¼ the known distance DG; we see also that DL, which is equal to AK, is very close in value to DE and AB. Similarly, CK is very close to BC, the length of the gaff. Let's stop here and solve the triangle as $AC^2 + AK^2 = BC^2$ (approximate). This may also be stated as BC equals the square root of the sum of AC^2 and AK^2.

In the situation just described, the desired dimension AB/DE lies within the

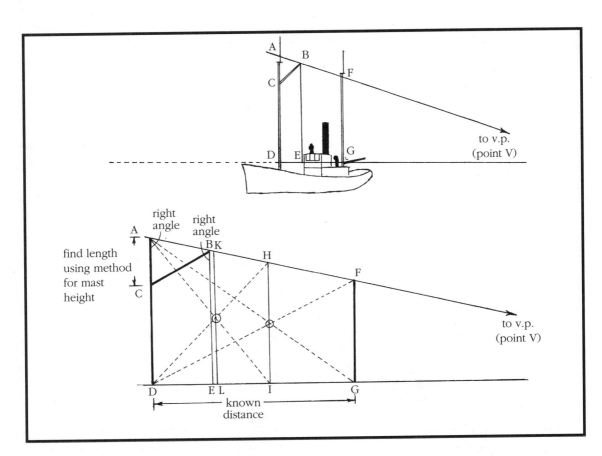

FIGURE 2-27. Finding the length of the gaff.

known dimension AF/DG. If the desired dimension lies outside a known dimension, as in the case of the mainmast boom in our example, there are a series of preliminary steps involved (refer to Figure 2-28).

Draw a line from point C where the boom intersects the mainmast to the right-hand vanishing point and extend the line to the left through the foremast at point D. You can use the same top line as in the previous exercise. Draw a vertical line through the aft end of the boom (point F), intersecting the reference lines at points E and G. Now we want to solve the triangle CFG for the length of the boom, CF. You can find FG in the same way you determined the length of the smokestack, so now you need to know the value of CG. Draw the diagonal AC, and

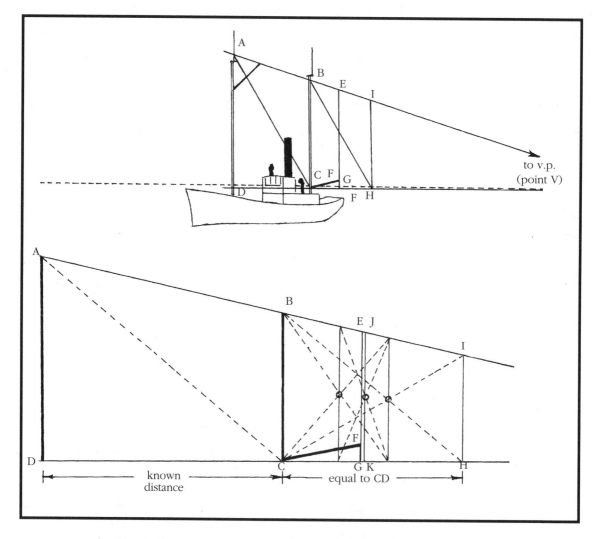

FIGURE 2-28. Finding the length of the boom.

then draw a line parallel to AC from point B to point H on the lower reference line. Drop a vertical through point H, intersecting the top reference line at point I. If you remember your Euclid, you will now see that the rectangle ABCD is identical to rectangle BIHC, so CH is equal to DC, which is a known distance. Now subdivide BIHC (three times in this case) as in the previous example to approximate CG. The final subdivision will give you the line JK, and CK is approximately equal to CG. Calculate the length of the boom by solving the equation $CK^2 + FG^2 = CF^2$ (approximate).

The foregoing is one example of the procedure for taking dimensions from photos or perspective drawings. I hope you haven't become suicidally confused. If you're interested in learning more about these techniques, consult an art book on perspective theory.

PLANNING THE WORK

When you finally have a set of drawings that satisfies you, it's time to do a little planning. Study the drawing from the standpoint of the materials required. What kinds of wood will be needed (covered in Chapter 3), and how much? At first, your estimates might be off, but with experience you'll get a feel for it. What fittings do you need, and how many of each? Will you buy them or make them? If you make them, what materials are required: brass sheet, wire and rod, special woods? How many blocks and deadeyes do you need? What sizes and what types? What about finishing materials: sealer, paints, stains, varnish? Are you going to make sails? How much cloth do you need? What about rigging line? Make up a bill of materials to the best of your ability.

Take a look at your tool set. Is there anything missing that you believe might be needed? Chapter 4 discusses tools you'll use in modeling.

Now give some consideration to the sequence of work. In later chapters there are many suggestions regarding this. Some things are obvious — you have to make the keel assembly before putting the frames on, and the deck beams have to be in place before you can plank the deck. But there are little traps. For example, mast hoops have to be put on the masts before the tops are assembled, and mast steps have to be built into the keelson at the time the keelson is installed. Some things can be done out of the normal sequence. If you get bored with planking, for example, take a break and make the mast assemblies or the sails. Just don't paint yourself into a corner.

CHAPTER 3

Materials

*"Apple, pear, cherry. . . ! This ain't a lumberyard,
it's a dadblamed orchard!"*

— Wilbur Woodsy, ship's carpenter

With kits, the materials are supplied, but you're on your own with scratchbuilding. The importance of using the correct materials in building a model cannot be overstated. The purpose of this chapter is to describe the properties of the most commonly used materials. See Appendix III for a listing of suppliers.

Fittings

A great many commercial model fittings of all kinds are available today. They range in quality from the exquisite to the grotesquely clumsy, but all have one thing in common: high price. Furthermore, in spite of the wide variety available, you might be unable to find exactly what you need. So what's the answer? Make them yourself. It's cheaper and more satisfying than if you bought them.

On the other hand, good commercial fittings are available, so don't hesitate to use them if they're right for your model. I don't make my own deadeyes or belaying pins, for example. There are fine ones available, and it's not worth the time to make them. I do make all my own blocks, however — many are needed, the price gets to be high, and they're not hard to make.

Make as many of the fittings as possible for your scratchbuilt model (otherwise it isn't really scratchbuilt, is it?). Never use commercial fittings that are ill-proportioned, out of scale, or inaccurate, no matter how convenient they might be.

GLUES

A bewildering array of glues is available today, not all of them suited to your purpose. Here are the most popular ones, listed from most to least useful.

CYANOACRYLIC (CA) GLUES

Cyanoacrylic glue (better known as CA glue) is a superior type of "super" glue. It comes in three types, classified according to curing time: The fastest cures in seconds and is meant for model airplane repairs on the flying field; the intermediate type takes up to two minutes to set, and is what I usually use; the slowest takes about five minutes, but is also quite useful. The intermediate and slow types are gap-filling glues. I have found CA glue most useful for planking. Because it sets up so quickly, there is no need for extensive, long-term clamping.

The bond with CA glue is indestructible; the glue literally welds wood together. A tiny drop on the shaft of an eyebolt before setting it into its hole ensures that the eyebolt is never going anywhere. It glues metal to metal, plastic to wood, anything to anything (including skin to skin, so be extremely careful). A solvent is available in case of emergencies, but don't wait for an emergency to happen. Get the solvent and store it in your workshop right next to the glue.

Now a few words of advice on using CA glue: Don't use it on rigging — it makes the line brittle. And if you get it on a wood surface that you intend to stain, forget it. That wood will never take stain. Don't try to glue clear styrene plastic with it either; it clouds the material. You don't even have to get the glue on the plastic — the fumes alone will do the damage.

ALIPHATIC RESIN

Popularly known as yellow carpenters' glue, aliphatic resin is sold under such trade names as Franklin's Titebond and Elmer's Carpenters' Glue. It's most useful for gluing wood and paper: It takes hold quickly, dries completely in less than an hour, and when clamped provides a strong, permanent bond. This is the glue I use most often.

AMBROID CEMENT

Ambroid is the trade name for a clear yellow cement that has been around for a long time. It's good for gluing wood to wood. I also use it for gluing small metal or plastic parts to wood when there will be no strain applied, and for coppering hulls. It's markedly superior to other clear cements.

PLASTIC CEMENTS

If you need to glue plastic to plastic, use the cements sold for building plastic models. The liquid type sold in small bottles is superior to the gummy type sold in tubes.

CONTACT CEMENT

Contact cements have limited applications in ship modeling. Many people believe that they're excellent for coppering, but I have doubts about their permanence. You can do without them.

ACETATE GLUE

Acetate glues such as Duco can be used for gluing wood and paper, though I don't consider them best for modeling. When thinned with acetone, however, they're good for temporary gluing — a touch of acetone releases the bond.

EPOXY CEMENT

Epoxy cements are two-part glues that must be mixed. They come in standard and quick-drying varieties, and their reputation as powerful, all-purpose cements is well deserved. The disadvantage is that you can't mix much at a time, so you're always stopping to mix another small batch.

HOT-MELT GLUE

Hot-melt glues — descendants of the classic glue pot on the stove — have become popular since modern formulas applied with an electric glue gun have become available. I find the glue gun cumbersome, though. Also, the heat from the gun can release a neighboring bond. This can be both inconvenient and frustrating when attempting to make a number of close, small joints.

MODEL AIRPLANE GLUE

Model airplane glue is meant for bonding balsa wood in model airplanes. It is not suitable for model ship building.

WHITE GLUE

The white glues, such as Elmer's, don't have the strength necessary for permanence in bonding wood.

METAL

Brass and copper are the metals used most by the model builder. Hobby shops stock brass in a wide array of shapes and sizes, including sheets, tubes, wire, and rods. If you need a brass rod over 3/16 inch in diameter, however, you'll probably have to go to a metal supply company. The main use for copper is for coppering the bottoms of hulls. For this you'll need sheets 0.001 to 0.003 inch thick. Try a hobby shop first, but you might have to go to a metal supply company.

You'll also need both hard and soft wire. Very fine wire has many uses, such as railings and stroppings for blocks. Fine wire is usually wound on spools and needs to be straightened before using. Grip one end in a vise and the other end in a pair of pliers and pull. This both stretches and straightens the wire.

RIGGING LINE

The best material for rigging lines is linen, but this is increasingly hard to find; check shipmodel supply houses. Common polyester/cotton thread sold in fabric stores, especially the type known as buttonhole twist, is good for the smaller sizes. Silk thread can also be used to advantage. Shop around to find what you need; you might even want to make up some line of your own (covered in Chapter 9).

Wood

Wood is the primary material in most models. Be choosy and select the proper wood for each application. Use only woods with a straight, close grain. If you plan to paint the model, though, there's no point in using expensive wood that will be hidden by the paint. Concentrate instead on how easy the wood is to work with and how it stands up over time. If you want a natural finish, choose woods that look good when varnished or oiled.

Unfortunately, it's getting harder to find many of the hardwoods. Some mail-order houses stock a few types, and doll-house stores usually have some on hand. Keep an eye out for trees that are being cut down in your area (you'll have to season such finds before using the wood, of course). Flea markets and junk shops are treasure troves of great old wood in the form of chairs, tables, and trunks.

Consider the following types of wood for your model:

Apple. An unusually hard wood, dark to light brown. Good for turning and for making small parts and spars.

Ash. Bends easily. Reported to be excellent for planking, though I have not tried it.

Basswood. An excellent all-around wood, readily available in hobby shops. White and very fine-grained. Easy to work with and bends freely, but too soft to turn. Needs to be carefully finished to avoid fuzziness, but can be stained to simulate many other woods.

Boxwood. Very hard, close-grained, and strong. Can be left natural or painted. The best wood for small carvings — holds minute details. Preferred material for block-making, and excellent for all small parts. Beautiful yellowish color improves with age.

Cherry. A hard wood with a dark brown color. An oiled finish is desirable. Can be bent, and is excellent for small parts; I have seen some very beautiful models made entirely from cherry.

Dogwood. Very hard, and excellent for turnings. Used for textile mill spindles because surface becomes harder and smoother with use.

Holly. A white, hard, close-grained wood. Excellent for turning. Can be used for small fittings, including blocks, and makes beautiful decks. Does not take stain well, though.

Mahogany. Hardness and closeness of grain varies by species, so be choosy. Takes a beautiful natural finish, and can be used in numerous applications.

Pear. Very hard and close-grained. Bends freely, and can be used for almost anything. Makes fine blocks. Takes a beautiful natural finish, and was much used by makers of British Admiralty models. Quite expensive.

Pine. Several varieties available: white pine is good, sugar pine is best. Sugar pine is an all-purpose, soft, white wood. Look for straight-grained pieces without knots. Very good for carving if fine detail is not a consideration. Should be painted.

Plywood. Use of plywood in ship modeling is limited, but it's useful at times.

Look for thin sheets sold for model airplane use.

Walnut. Hard, with beautiful dark natural finish. Insist on close, even grain. Can be bent, but great care is required. Makes very nice fittings and desk furnishings.

MISCELLANEOUS MATERIALS

Your home workshop is not complete without a few other interesting materials.

ACRYLIC PASTE

Sold in art supply stores, acrylic paste is not a glue, but rather a thick white paste that dries into a hard material that you can easily carve and sand. Since it holds detail well, I use it for modeling complex shapes such as crew members (covered in Chapter 11).

GRAPHICS TAPE

Self-adhesive graphics tape, available in drawing supply stores, comes in a wide range of colors and widths starting at $1/64$ inch. Use it for making mast and anchor stock bands.

PAPER

Card stock (such as index cards or smooth, somewhat heavier Bristol board) is a must-have. (Acid-free stock is best, but may be hard to find.) Because the stock has no grain, it's ideal for small, thin pieces such as window frames and paneled doors. It's almost impossible to cut these out of any kind of wood without cracking it, but card stock does the job nicely. Card stock is also indispensable for making tiny windows or for simulating thin sheet metal in models of small-scale modern ships. Fragile? Not if you enhance its strength with a coat of thin varnish on both sides before using it. The varnish soaks in and binds the fibers. You can sand the finished product, or laminate two or more layers to build up thickness. It's worth trying, and you'll find many uses for it. The bridge structure illustrated in Figure 3-1, for example, is made entirely of index card stock.

PLASTICS

Many "serious" modelers unjustifiably look down their noses at plastics. If used appropriately, though, it's a wonderful material, even on the finest models. Sheet plastic is readily available in a variety of thicknesses and can serve many of the same purposes as paper. It's somewhat more difficult to work with than paper, especially if it's fairly thick. And like paper, it must be painted. Clear polystyrene plastic is the best material for simulating glass. When choosing a thickness, be sure the plastic is thick enough to be rigid when cut to size.

PLASTIC WOOD

Plastic woods have come a long way. The best are products such as epoxy putty that are activated by a catalyst. Look for these products in woodcarvers' supply catalogs. When hard, they carve beautifully, and can be used in much the same way as acrylic paste. If you cut too deeply and need a

FIGURE 3-1. The bridge structure and boat cranes on this model were made entirely of index card stock. (Photo by Patricia Leaf)

wood filler, this is the thing to use. It's expensive, but you'll find many uses for it.

POLYESTER CASTING RESIN

Although polyester casting resin doesn't have a great many uses in ship modeling, it's invaluable at times, such as when casting certain small parts. I've used it to make swimming pools on ocean liners, and to create the "glass" in airports and searchlights (covered in Chapter 11).

TULLE

Dressmakers' tulle is a stiff, fine net that you might find in better fabric shops. Use it for the top nettings on the fighting tops of sailing warships, for hammock rails, for the safety nets often installed under the bowsprit and jib booms, and even for the torpedo nettings used on World War I–era capital ships.

CHAPTER 4

Tools

"A workman is only as good as his tools."

— Anonymous savant

Tools are personal things. Over time one accumulates all sorts of things believed to be useful. Some really are, some are relegated to the back of the drawer, and some are discarded only to become another person's favorites. Here are the tools you'll really use, but add your own personal favorites, if you like.

DRAFTING EQUIPMENT

Scratchbuilding requires a few pieces of drafting equipment. For starters, you'll need a drawing board, a T-square or parallel straightedge, and triangles (or, if you can afford it, a drafting machine, which combines the functions of the T-square and triangles). You should also consider the following:

- A standard set of drawing instruments, which includes both small and large dividers and compasses. The dividers are for transferring measurements directly, and the compasses are for drawing circles and arcs.
- A pair of proportional dividers. These are double-ended dividers with a sliding pivot that allows you to change the ratio between the two ends. By setting the ratio to 1:3, for example, any mea-

surement taken at the small end will automatically be enlarged three times at the large end. These dividers are handy when enlarging or reducing drawings.

- A protractor with a radius of 4 to 6 inches for accurately measuring angles.
- Architects' and engineers' scales. These traditionally three-sided rulers have six different scales, one of which is a standard ruler (scale of 1:1). The others read scales directly — for example, 2:1 or $\frac{3}{16}''=1'$. Architects' scales are graduated in feet and inches; engineers' scales are graduated in feet and tenths of feet.
- A set of "ship curves" (available in good drawing supply stores). These are similar to the traditional French curves, but are designed especially for ship drafting.
- A small pair of outside calipers used for doing turnings on a lathe.
- A plastic spline that can be set to any desired curve. This is a particularly useful item that I recommend highly.
- A light box for making tracings.
- Drawing pencils of varying degrees of hardness. 3B is the softest you'll need, and 5H the hardest.
- A supply of paper, including tracing paper. Some people like to use graph paper, but I find that it confuses my own grids.
- Drafting tape to secure paper to the drawing board. It's much better than using drawing pins, which can tear the paper slightly and dislocate it.
- A good eraser for the mistakes even the best of us make all too frequently.

WOODCARVING TOOLS

Look in a catalog of woodcarving tools and you'll be overwhelmed by the different types and sizes available. You really don't need many: a flat chisel, an angled chisel, a curved chisel, a V-gouge, and a U-gouge. A set of miniature tools is also invaluable for carving figureheads and transoms. If you build lift models, you'll need larger tools.

Planes, spokeshaves, and drawknives are used primarily in making lift models. They come in an assortment of shapes and sizes. Look for them in a good tool store or a woodcarvers' catalog.

My favorite knife is an X-acto with a #11 blade. This small, sharp, triangular blade is what I use 90 percent of the time; only occasionally do I need a larger blade. Choose your own knives; time will tell what you actually find most useful.

COMMON TOOLS

Pliers. Several kinds of small pliers are essential, including needle-nose, round nose, wire-cutter, and utility pliers.

Drill bits. Be sure to have a set of #61 to #80 drill bits, and some larger sizes as well, up to ¼ inch at least. Some you won't need too often, but they will be there when you do. Get a **pin vise** to hold the smaller sizes and a hand drill for the larger ones. You can also use them in a rotary power tool or drill press if you have one.

Clamps. An assortment of small C-clamps is essential. You'll need larger ones if you plan to build lift models. X-acto sells small tweezer-like clamps that are handy, and common spring clothespins are invaluable. Pipe cleaners and the wire twist ties used to fasten plastic bags can secure planking battens to the frames. Small tweezers, both with straight and bent ends, are also a must. Consider a pair of long tweezers (8 to

10 inches), a pair of self-closing ones, and a couple of surgical hemostats — great for locking a grip on something.

Files and rifflers. These are invaluable in shaping both wood and metal. Be sure to get flat and round files in varying degrees of coarseness. Sets of small files and the specially shaped rifflers are available in hobby shops. I also occasionally use larger ones with 6- to 8-inch blades.

Saws. The most useful saws I've ever used for modeling are the large and small X-acto razor saws, and the X-acto saw blade #13. Jewelers' saws are favored by many. A small miter box is available to use with the razor saws. You might find similar items that you prefer.

Square. A 6-inch machinists' square is good for most purposes, but a 12-inch adjustable square is great for when you need to mark up large pieces of wood. An adjustable square enables you to mark angles other than 90 degrees.

Vise. You should have a small all-purpose vise, and if you're going to build lift models, you should also have a full-size woodworking vise.

POWER TOOLS

Your investment in power tools depends on the size of your wallet. I got along for a long time with only a Dremel tool (a hand-held rotary power drill/router/etc. holder), a small assortment of accessories (cut-off discs, small circular saws, and a few variously shaped routers), and a small jigsaw with attached sander. I eventually acquired a simple wood-working lathe and a proper sanding machine. I honestly don't know how I lived without the sanding machine. Then I discovered a small orbital sander, a wonderful tool that saves hours of tedious hand sanding. Eventually I got a good bandsaw, and then wished I had one years earlier. I prefer it to a jig or fret saw. My latest acquisition was a Unimat metal-working lathe to which I am still adding accessories. This is an expensive precision item, but well worth the money. It can be used for turning, milling, as a drill press or table saw, and much more. A similar machine is made by Sherline and is said to be very good.

SPECIALIZED TOOLS

Plank bender. This is one item I don't use, but it's so popular it's worth mentioning. Electric plank benders pass a strip of wood through heated rollers to bend it, and they do a good job indeed.

Rigging tools. The single most useful rigging tool is an ordinary needle threader, found in sewing notions departments. Use them for threading lines through blocks. They don't last long, but they're cheap, so keep a good supply on hand. I once ordered a great set of rigging tools from a catalog. They're ingenious, consisting of large needles set into dowels with the eyes cut and bent into different shapes. Unfortunately, they're too big for many uses, so I made some smaller ones using common sewing needles. Figure 4-1 shows some useful patterns. You can use the cut-off disk in your power tool to cut the shapes, and heat the needles to make the bends.

Rope-making machine. If you decide to make up your own line, you'll need a rope-making machine. I use a simple hand-operated machine — the only one I've ever seen on the market — but I've seen some remarkable home-built powered machines. The Nautical Research Guild has published articles about making such machines; see Chapter 9 for more on this subject.

Miscellaneous Tools

Glue syringe. Hobby shops stock glue syringes that enable you to place small amounts of glue precisely. Some are double-barreled, automatically mixing two-part epoxy cements. I use them occasionally, but don't recommend them for use with CA glues.

Rulers. A 12-inch steel machinists' rule is most useful, and a metal yardstick to use as a straightedge is essential.

Scissors. I have a pair of good standard sewing scissors for cutting patterns, sails, and the like, a couple of pairs of nail and cuticle scissors for rigging work, and a small pair of metal shears.

Soldering iron. Some modelers swear by a 25-watt conventional soldering iron, while others prefer a soldering gun.

"Third Hand." A "third hand" is a device with alligator clips mounted on adjustable arms. It holds an item in place, leaving both your hands free. It's great for jobs like soldering small pieces together and holding line when seizing.

Toothpicks. You'll find that a supply

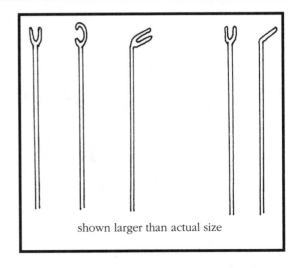

shown larger than actual size

FIGURE 4-1. Rigging tools made from needles.

of flat toothpicks is indispensable, especially for applying small amounts of glue.

Sandpaper. You need an assortment of sandpapers from #60 to #600. Except for the coarser grades, wet-or-dry papers are best. I also keep on hand a can of pumice powder for final finishing of paint jobs.

Beeswax. A block of beeswax is needed for waxing rigging line.

Paints and brushes. Sable brushes last long enough to justify their price (if you take care of them), and inferior ones are frustrating to use and give poor results. You'll need both rounds and flats in a range of sizes. Flats are best for covering large areas; small rounds are best for detailing. Buy good quality model paints. Chapter 8 includes a discussion of suitable paints.

CHAPTER 5

Half-Models

"At first I couldn't understand this revolutionary design — until I realized that Mr. McKlutz had literally copied the half-model!"

—James Eyeball, ship critic

If you've never scratchbuilt before, you might want to try a half-model before embarking on a full-hull, rigged model. Half-model construction is easily controlled, and it can stand on its own as a display item. You can make a half-model of a ship you intend to model completely later, or of something quite different. In either case, you'll gain valuable experience.

Half-models were once used as design aids by ship designers and shipwrights in addition to, or instead of, drawings. They were invaluable to shipbuilders who lacked the expertise to make accurate drawings. A common way to use them was to bend lead bars over the model at the stations and then to transfer the resulting shapes to the lofting floor. For accuracy, such models were usually made to a large scale, and they were very plain. Today, original half-models of this type have considerable value to collectors. Until recently, even with the availability of sophisticated drafting techniques, half-models were used to lay out the plating on steel ships. A wonderful model of this type, of the liner *United States*, is exhibited at the Mariner's Museum in Newport News, Virginia.

In time, the idea of using half-models purely for display became popular, and the models became more elaborate, with rudder, stub masts and bowsprit, deck furniture, and painted or highly finished natural wood. One reason for their popularity is ease of display. They simply hang on the wall like paintings. Half-models are satisfying to build and, in cases where only the lines drawing of a particular ship has survived, might be the only good way to model the ship.

Hull construction

There are four ways to build a half-model hull: solid block, lift, hawk's nest, and plank-on-bulkhead.

Solid-Block Hull

Step One. Mark the shape of the deck (from the plan view) on the block and bandsaw the block to this shape (**A** and **B** in Figure 5-1). I recommend basswood for a first attempt at this type of model, but you can use any close-grained wood.

Step Two. Replace the piece you cut away with a few spots of rubber cement so you'll have a flat surface on which to mark the profile (the next cut), and so that the block will lie flat on the cutting table (**C** in Figure 5-1). (Using rubber cement in this application enables you to later break the seal easily.)

Step Three. Mark the profile on the block, omitting the cutwater, keel, and sternpost shapes.

Step Four. Cut out the profile and dispose of all the scrap pieces. You now have a three-dimensional outline of the hull ready for carving (**D** and **E** in Figure 5-1).

Step Five. Mark the positions of the sections on the flat back of the block and prepare templates, using the shapes shown in the section view of the drawing. See Figure 6-28 in the next chapter for the use of the template.

Step Six. Carve the block to the shape of the sections, using the templates frequently. This is a bit tricky, since there is no inherent guide. Rough-carve the hull to the approximate shape, being careful not to cut too deeply. Establish a smooth transition between the sections. Now refine the cuts until the templates show conformity to the required section shapes. Finish with sandpaper (**F** in Figure 5-1).

Step Seven. Make the cutwater, keel, and sternpost, and glue them to the hull (Figure 5-2). You can also add bulwarks, if applicable. If you do this, you'll have to mortise timberheads into the hull and plank over them to form the bulwarks. Refer to Figure 6-29 in the next chapter for the mortising technique.

Lift Hull

A lift model of a half-hull is constructed in the same way as you would a full-hull lift model (covered under the heading "Lift Method" in Chapter 6). This type of construction is admirably suited to fine woods and a natural finish. You can, for example, use alternating light and dark woods for the lifts.

Hawk's Nest Hull

Hawk's nest models were originally used to check the frame shapes for fair planking. Although few models of this type

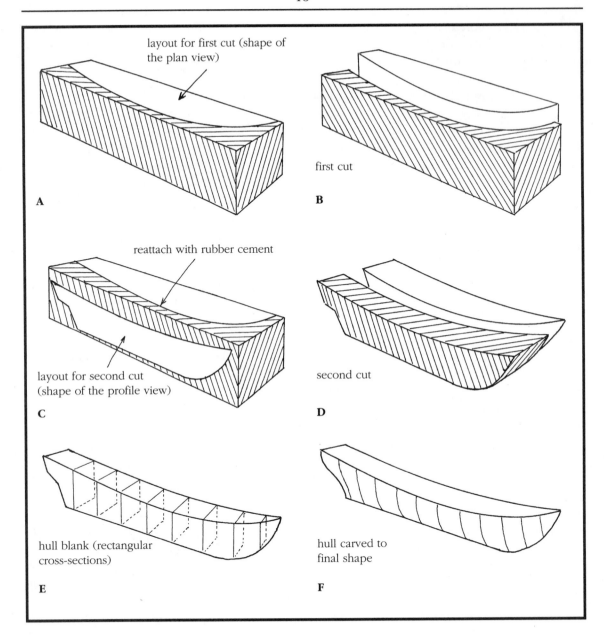

layout for first cut (shape of the plan view)

A

first cut

B

reattach with rubber cement

layout for second cut (shape of the profile view)

C

second cut

D

hull blank (rectangular cross-sections)

E

hull carved to final shape

F

FIGURE 5-1.   Shaping the hull for a solid-block half-model.

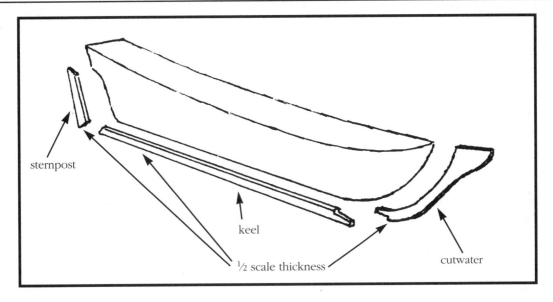

FIGURE 5-2.   Adding details to the solid-block hull.

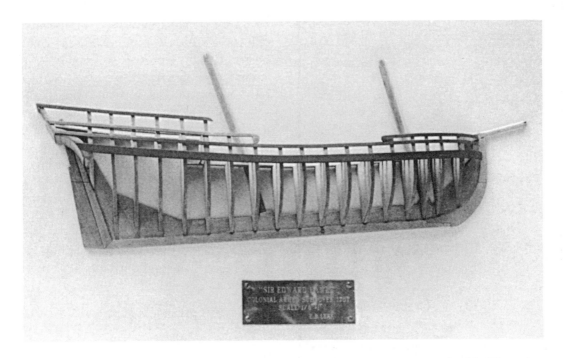

FIGURE 5-3.   Hawk's nest half-models are unusual and most interesting to build. (Photo by Patricia Leaf)

are being made today, they're most interesting and striking in appearance (Figure 5-3).

Step One. Cut out a base piece to the shape of the profile; include the cutwater, keel, and sternpost (**A** in Figure 5-4). This piece should be *half* the scale thickness of the keel.

Step Two. Cut the keel, stem, and sternpost rabbets into the base piece.

Step Three. Cut out half-bulkheads (frames) — including false timberheads for the bulwarks and rails — and glue them to the base piece (**B** in Figure 5-4).

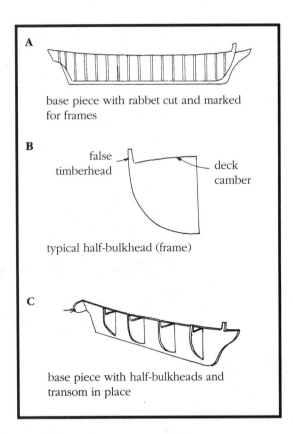

base piece with rabbet cut and marked for frames

false timberhead   deck camber

typical half-bulkhead (frame)

base piece with half-bulkheads and transom in place

FIGURE 5-4. Construction details for a hawk's nest half-model.

Step Four. Add the transom (**C** in Figure 5-4).

Step Five. Now add thin strips of wood at the positions of the plank sheer, wales, and rail.

Step Six. If you wish, you can now add cap rails and a few deck planks. The only other appropriate details would be a rudder, stub masts, and a stub bowsprit.

Hawk's nest models look best with a natural finish. For added interest, you can use different woods for the base piece, bulkheads, and planking.

VARIATIONS ON THE HAWK'S NEST MODEL. A nice variation to the solid base piece is to substitute a properly shaped representation of the keel, keelson, stem, sternpost, and deadwood (Figure 5-5).

Another variation is to use fully shaped half-frames and deck beams instead of bulkheads. This is elaborate, and while not necessary in capturing the essence of the old hawk's nest models, is nice to do.

The keel is notched to receive half-frames in lieu of bulkheads. You'll have to use half deck beams to support the half-frames. Since this method of construction doesn't provide as much control as is provided with the solid base piece, you might find it convenient to build the model directly on the mounting board. If you do, be extremely careful to keep the mounting board from being damaged during construction.

PLANK-ON-BULKHEAD HULL

A plank-on-bulkhead half-model is an extension of the hawk's nest model. Simply build a hawk's nest model, but give it full planking. If you don't like to carve, this is a

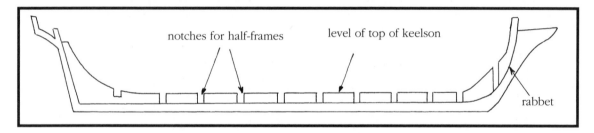

Figure 5-5. Keel assembly for a modified hawk's nest half-model.

great way to build a half-model. And if you've never planked, the plank-on-bulkhead model is a good way to start since it's easy to align the frames on the base piece.

ADDING DECORATION

Convention has it that any half-model that includes more than the basic hull is a *decorated* model. The first things to add are the rudder, stub bowsprit, and stub masts. From that point you can get as elaborate as you wish, adding deck furniture and fittings, channels and lower deadeyes, catheads and anchors, figurehead and other carvings, and so on. Obviously, many of these details will be half-modeled. There should be no rigging of any kind.

DISPLAYING YOUR HALF-MODEL

The backboard on which the model is mounted is the final item. Although some modelers simply insert screw eyes in the top of the model for hanging without a backboard, this seems austere. A backboard of good wood set off by a simple frame is best. You can attach the model to the board with screws entering from the back of the backboard and into the back of the model. I've even seen old decorated half-models mounted on a mirror so as to give the illusion of a full hull model. It's an interesting idea, and you could do this by using CA glue to cement the model to the mirror.

CHAPTER 6

Hull Construction

*"I really likes t'build planked hulls — them solid
ones busts up the marine railway."*

— Joshua McKlutz, progressive shipbuilder

PLANKED HULLS

The two most common methods of planked
hulls are plank-on-frame and plank-on-
bulkhead. Related to this is the construction
of decks, which is done in the same way for
both types of planked hulls.

PLANK-ON-FRAME HULL

In its pure form, plank-on-frame construction is an attempt to reproduce as closely as
possible the structure of a real ship. Methods for building the framework are well described in H. Hahn's article in the Nautical
Research Guild's *Ship Modeler's Shop Notes*
and in H. Underhill's *Plank-on-Frame Models* (refer to Appendix II for publishing information). The work must be done from a
drawing to the inside of the planking. You
begin just as you would for a real ship — by
building the keel assembly.

BUILDING THE KEEL ASSEMBLY. First, decide whether to build the keel assembly using a keelson (a heavy timber that rests on
top of the frames, locking them to the keel),
or to make a one-piece deep keel whose
height is equal to the distance from the bot-

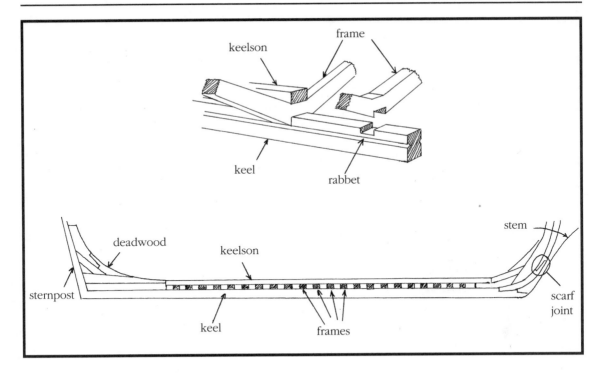

FIGURE 6-1. Assembling the keel.

tom of the keel to the top of what would be the keelson. If you choose to use a keelson, your assembly will look like the one in Figure 6-1. If you choose the deep keel, it will look like the one in Figure 6-2. The deep keel is easier for the beginner.

Whichever choice you make, you'll have to do the rabbets, into which the garboard edges and the plank ends fit. There are two ways to do this. The first, which replicates actual practice, is to cut the rabbets into the keel assembly, as has been done in Figure 6-1. The alternative, which might be easier for some modelers, is to build the keel assembly in three layers, as shown in Figure 6-3.

When you have made your decision, you can proceed with construction.

Step One. Depending on your decision:

- Build up the keel assembly to full thickness and cut the rabbets. If you have chosen to use a keelson, cut shallow notches in the top of the keel at the locations of the frames, as shown in Figure 6-1. Otherwise, cut deep notches, as in Figure 6-2.
- Or, build the keel assembly in three layers (Figure 6-3): a thick central core the full shape of the assembly, with thin side pieces shaped only to the outer rabbet. The total thickness equals the correct thickness of the keel. The thickness of the side pieces depends on the thickness of the planking you intend to use. For example, if you plank with ¹⁄₁₆-

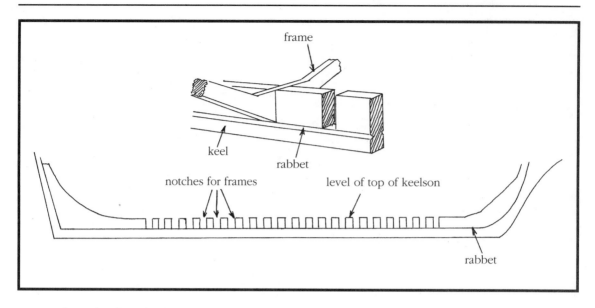

FIGURE 6-2. The deep-keel option.

inch stock, use ¹⁄₁₆-inch material for the side pieces. Be sure that the layers are well glued, and use plenty of clamps. Although this method of building the keel assembly isn't true to life, as is the first method, it's acceptable. You can use either the keelson or the deep keel with this method.

Step Two. Now you make the keelson, if you decide to use one, even though you won't install it until later. Build the mast steps either into the keelson (or the top of the deep keel) or into the deadwood, depending on their locations as shown on your drawing. The tenons at the bases of the lower masts will fit into these steps.

Step Three. Build up the frames (Figure 6-4). These are made in two layers, each made up of several pieces arranged so that

the grain of the wood follows the shape of the frame as nearly as possible, and so that the pieces in one layer overlap those in the other layer. If you are using the keelson option, cut shallow notches in the bottoms of the frames to match the notches already cut in the keel, as shown in Figure 6-1. These notches are not necessary for the deep-keel option (Figure 6-2). In both cases, the outer surfaces of the frames must intersect the keel assembly at the level of the inner rabbet or of the bearding line.

Step Four. Bevel the frames so the planking will lie smoothly on them. Although you can do the beveling before attaching the frames to the keel, many modelers find it easier to do the beveling after the framework is assembled. Pre-beveling is helpful, however, for frames near the bow and stern that might have considerable and complex bevels.

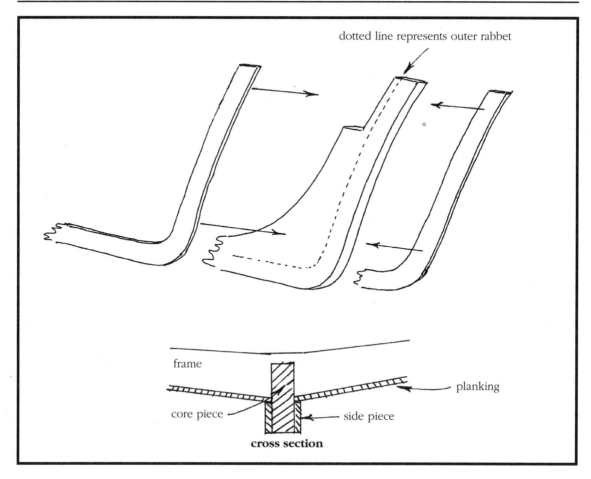

dotted line represents outer rabbet

frame

planking

core piece

side piece

cross section

FIGURE 6-3. Another way to build up the keel.

Step Five. Clamp the assembled keel, stem, and sternpost into a jig. Supporting the framework during assembly is the key to an accurate hull. Hahn's method provides great control. I've managed very well with the clamping jig shown in Figure 6-5, and the frame aligning jig shown in Figure 6-6.

The clamping jig is constructed so that it can be used for many models. Mine is a good flat pine plank about 4 feet long, 8 inches wide, and ¾ inch thick. Mark a permanent centerline down the face of the plank. The clamps are made from metal angles faced with wood and screwed to the plank in pairs. The pairs are spaced along the centerline so that they can clamp the stem and sternposts. The transverse spacing between the two sides of each pair should be equal to the thickness of the keel assembly. Small blocks of wood are tacked on either side of the centerline to keep the keel straight. The clamps and blocks can be

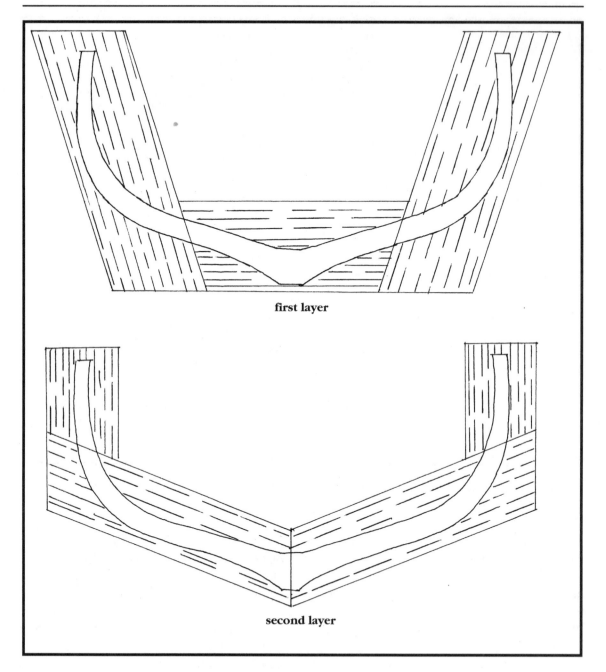

first layer

second layer

FIGURE 6-4. The frames are assembled in two layers, each consisting of several futtock pieces.

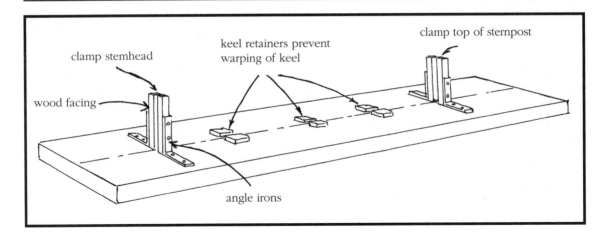

FIGURE 6-5. The clamping jig holds the keel assembly rigid during construction.

repositioned depending on the size of the model currently under construction.

The aligning jig can be cut from ¹⁄₁₆- to ¹⁄₈-inch plywood or, in the case of small models, stiff cardboard. The criterion is that it be rigid enough to maintain its shape.

Step Six. Place the aligning jig over the tops of the stem and sternposts, and set the frames into the keel with their timberheads in the slots of the jig.

Step Seven. When all frames are in place, but before removing the aligning jig,

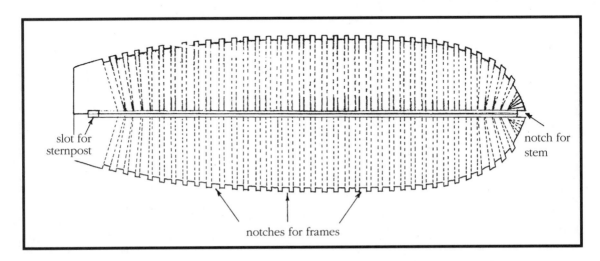

FIGURE 6-6. The aligning jig ensures that the frames are held in the proper position during assembly of the ship's skeleton.

glue small pieces of wood between the frames just below the level of the deck to preserve the spacing of the timberheads.

Step Eight. Remove the aligning jig and install the keelson (if you're using one), beam clamps and shelves, bilge stringers, and limber strakes. (The beam clamps and shelves are longitudinal planks inside the hull on which the deck beams rest. Unless you plan to leave part of the deck unplanked to show construction details, you can omit the beam clamp.) Refer to Figure 6-7 for the relative positions of these pieces.

Installing these pieces requires careful measuring, since the top of the beam shelves must be at the level of the bottom of the deck beams. The bilge stringers and limber strakes are optional, but may be used to further strengthen the hull.

FRAMING THE DECK. Accurate framing under the deck is critical for a plank-on-

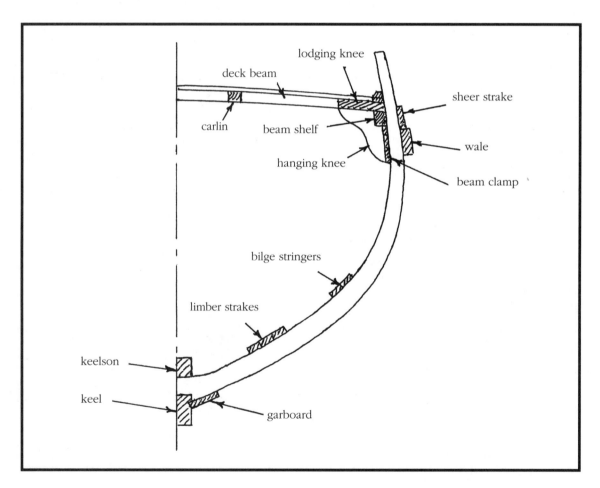

FIGURE 6-7. The reinforcing timbers are essential to the strength of a real ship, but you might not need all of them for a model.

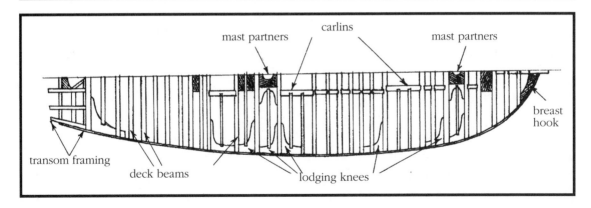

FIGURE 6-8. If any portion of the deck framing is to be exposed, you'll want to include all the details.

frame model. Framing the deck just like the real ship (Figure 6-8) is ideal, but this is unnecessary work unless you plan to leave part of the framing exposed, as in an Admiralty model.

You won't always find a drawing that provides the level of detail shown in Figure 6-8. Deck beams are normally about as wide as the frames are thick, and are square in cross-section. They are also cambered — curved so that water will not pool on the deck. The camber is a sector of a circle, and a safe curvature is ¼ inch for each foot of chord. For example, if the beam of the ship is 32 feet at midsection, the camber would be 8 inches at the centerline. Make a pattern for the midsection, and use it for all other beams, regardless of length, so that they all have the same curvature. The beams must be cut and installed carefully so that the deck planking lies evenly, just like the planking of the hull.

There should not be more than 1 inch between deck beams, and if you're working from an actual plan, they might be closer (depending on the scale). More often than not, the beam locations won't coincide with the frames. Spacing and loca-

tion are dictated by such considerations as deck openings.

In addition to the beams, you'll also need to build in the carlins that frame the hatches and other deck openings, and the partners for the masts. Be careful when positioning the mast partners. They must be correctly placed in relation to the mast steps so that the masts, passing through them and standing in the mast steps, will stand at the angles relative to the deck, as shown in the drawing. I always make the lower masts at this point, and use them to check the adjustment of the partners.

On real ships, not all deck beams are the same size — some heavier beams are used for lateral strength. If you plan to plank the deck completely, you can ignore this detail. Likewise, you can omit the hanging and lodging knees. Just be sure to include these items if you leave the deck unplanked. Once the deck framing is complete, the framework will be quite rigid and you'll be ready to proceed with planking.

PLANKING THE HULL. **Step One.** Make a working cradle (Figure 6-9). You're ready to remove the hull from the jig, and you don't

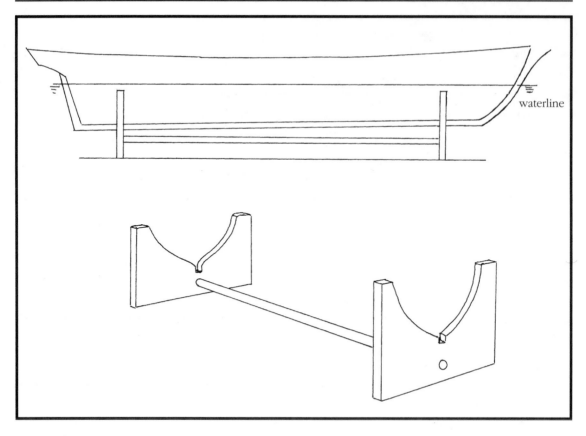

FIGURE 6-9. A working cradle protects your model during construction.

want it rolling around the workbench. Make the cradle so that it holds the hull steady with the waterline level (this will come in handy when it's time to mark the waterline).

Step Two. Now you need to determine the runs of the planking. You can do this by using a thin strip of paper to divide the edge of each frame into as many equal parts as there will be planks at the midsection. Proportional dividers may also be used by adjusting them to the ratio 1:number of planks. The number of planks will be determined by their width at the midsection and the length of one edge of the midsection frame.

Plank width varied in actual practice depending on available materials, but 8- to 12-inch scale width is reasonable.

When you have determined the number of planks, lay a strip along each frame to determine the length of the edge, straighten the tape, divide it into as many equal parts as required by the number of planks, and then use the tape to mark the frame.

Next, lay narrow strips of tracing paper along the hull, tick off the pairs of points on each frame that represent a plank width, and connect the points to determine the plank shapes.

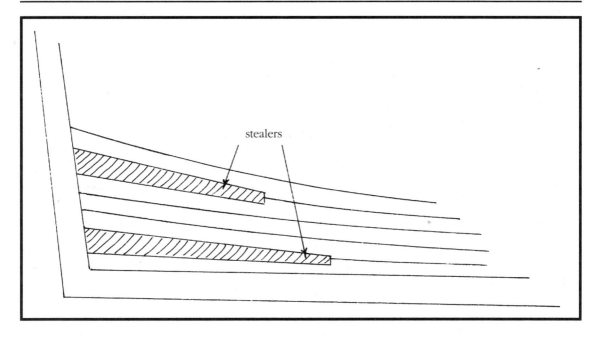

FIGURE 6-10. Stern stealers fill in gaps caused by the natural run of the planks.

The resulting shapes, however, might not be realistic because the method doesn't take into consideration the need for stealers at the stern and bow. At the stern, the plank shapes will probably flare to widths greater than that of the planks you'll use; and at the bow, the planks might become impractically narrow. To compensate for this, you need to add stealers — short planks of the same thickness as the others (Figures 6-10 and 6-11). At the stern, stealers compensate for the plank flare, filling the gaps left by the

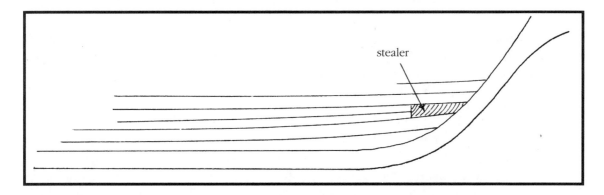

FIGURE 6-11. Bow stealers compensate when the forward ends of planks become too narrow.

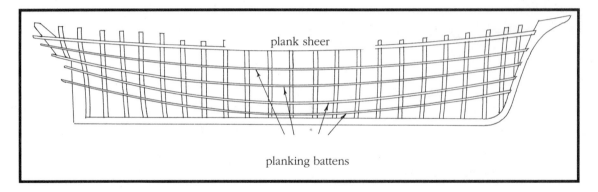

FIGURE 6-12. Planking battens can be used as guides in establishing the run of the planking.

standard planks. At the bow, they take the place of unacceptably narrow plank ends.

Another way to determine the run of planking is to use planking battens. These are thin, temporary strips of wood that you pin or tie to the frames to define the basic planking pattern. Refer to Figure 6-12. The garboard and sheer strakes provide the outer indices for the battens. This technique requires a certain amount of experience in planking so that you can identify reasonable runs. It is not recommended for a first or second effort.

Step Three. Now you're ready to do the actual planking. A sequence I've used successfully is to first install the sheer and garboard strakes. (The sheer strake is the longitudinal plank at the level of the main deck, though it doesn't always follow the line of the deck; the garboard strake is the strake adjacent to the keel.)

Step Four. When the garboard strake is in place, lay a long, straight plank along its edge. As you bend this plank along the frames, it will partially overlap the garboard at the bow, so you'll have to trim the garboard to fit. At the stern, the plank will

probably not overlap, and might even gap enough to require a stealer.

Step Five. When you have a fit, install the plank, full width. Repeat the process until the hull is planked. Areas requiring stealers will show up readily.

Step Six. Use CA or yellow carpenters' glue to fasten the planks both to the frames and to each other. Clamp everything in place. CA glue is ideal for this application because it dries quickly. If you plan a natural finish for the hull, however, you can't use the CA glue. Don't use little nails or pins for extra holding power; modern glues don't need them, and they tend to work loose anyway. You may, however, use treenails — wooden pegs set into holes bored through the planks and into the frames and then cut off flush. Bamboo is the material of choice for treenails. Simply draw the bamboo through successively smaller holes in a steel plate until you obtain the proper diameter. Just don't make them too big or your model will look as though it has the measles. Done judiciously, treenails add to the beauty of the model. If you plan to paint the model, how-

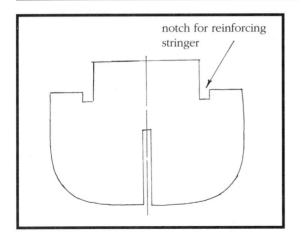

FIGURE 6-13. Bulkheads can be used in lieu of built-up frames.

ever, treenails are a waste of time because they won't show.

PLANK-ON-BULKHEAD HULL

The other method of planked hull construction uses shaped bulkheads (Figure 6-13) instead of frames. This method is most often used for models of steel ships, such as modern warships and merchantmen, though it can also be used as a shortcut in building wooden ship hulls. Don't use thin plywood for this method, and use as many bulkheads as you would frames.

PREPARING THE SKELETON. **Step One.** An easy way to construct the skeleton for a modern ship is to start with a solid backbone and notch it to receive notched bulkheads. (Figures 6-14 and 6-15).

Step Two. Reinforce the structure with longitudinal stringers set into notches in the tops of the bulkheads. (Unless the model is large — say, 3 feet or more in length and more than 6 inches wide — one stringer on each side is sufficient.) Note that the distance between the bottom of the bulkheads and the bottom of the backbone is the thickness of your planking material. You can extend the bulkheads to the upper decks, if you like. This has been done in Figure 6-13.

PLANKING THE HULL. For the most part, planking over bulkheads is the same as planking over framing, with the following exceptions.

The bows and sterns of most steel ships are difficult, if not impossible, to plank. Instead, plank only to the foremost and aftmost bulkheads. Then insert blocks of wood from the foremost bulkhead to the bow and from the aftermost bulkhead to the stern. Carve down the blocks of wood to the planking (Figure 6-16).

A better, but more difficult, way to do this is to first insert the blocks, carve them

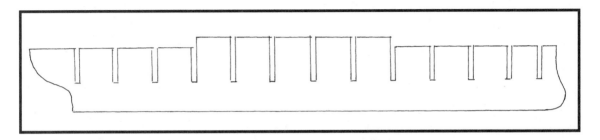

FIGURE 6-14. The backbone takes the place of the keel assembly.

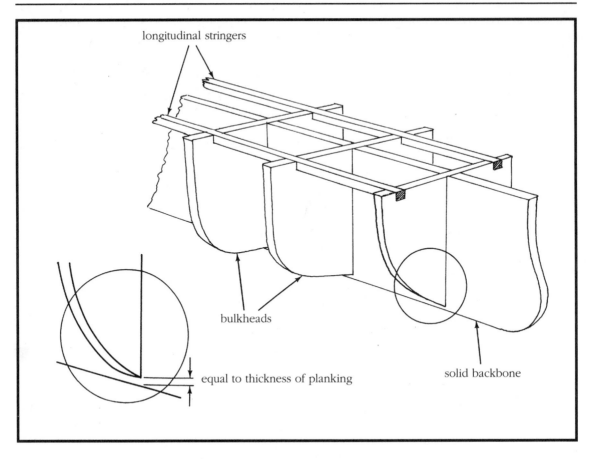

longitudinal stringers

bulkheads

equal to thickness of planking

solid backbone

FIGURE 6-15. Assembling the backbone and the bulkheads.

down to the bulkheads, and then let the planking extend over the blocks, after which you can carve them to their final shape. This method has the advantages of covering the seams between the plank ends and the blocks, and adds strength and durability to the model.

Because many modern ships have rather flat bottoms, the equivalent of the garboard strake might need to be very wide. If you're faced with this problem, build up the garboard with parallel planks as shown in Figure 6-17. Use a batten to determine the best shape for the bottom structure. When

the bottom structure is finished, continue planking in the usual way.

PLANKING THE DECK

Decks are done in the same way for both plank-on-bulkhead and plank-on-frame models. They look best when laid up with individual planks. Never scribe deck planks. Even if you have a carved hull, lay in the margin planks, waterways, and deck planks (Figure 6-18).

Plank ends should never come to a sharp point, so you'll have to fit the ends of

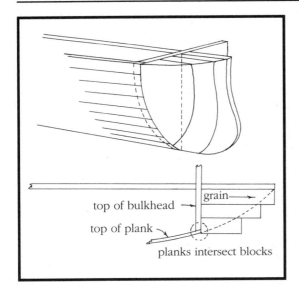

FIGURE 6-16. The bow and stern blocks are made up of diminishing laminations to simplify carving.

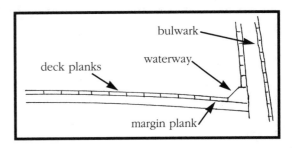

FIGURE 6-18. A typical cross-section at the waterway.

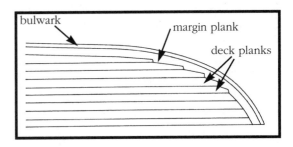

FIGURE 6-19. Deck plank nibbing not only looks right, but it makes it easier to fit the planks to the margin plank.

the planks into the margin plank or nibbing strake (Figure 6-19). (A distinction used to be made between the terms *nibbing* and *joggling*: Nibbing referred to wooden construction, and joggling to steel construction. Today the terms tend to be used interchangeably.) A good rule is to use nibbing when the angle at the end of the plank where it meets the margin plank is less than 45 degrees.

So that the caulked seams of the deck show up well, paint one edge and both

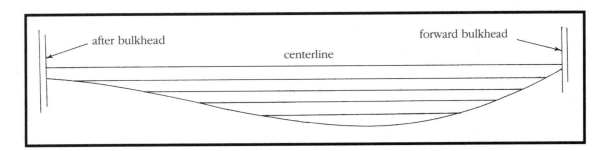

FIGURE 6-17. The bottoms of most modern ships do not lend themselves to conventional garboard construction.

ends of each plank with flat black paint. Lay the planks with a painted edge against an unpainted edge. After you lay the entire deck and sand it, the paint will show up as fine black lines. Some modelers glue black paper to the plank edges to achieve this effect, but it's a tedious and difficult process that works well only on large-scale models.

Don't use a single plank for the entire length of the deck — it's not realistic. Planks on real ships are usually no longer than 12 feet (depending on what was available at the time), so scale down the length of your planks accordingly.

Since the plank ends must rest on a beam, the actual length is, to some extent, determined by the spacing of the deck beams. Also, stagger the ends so that two

butt joints are not adjacent to each other. For small-scale models, such as $\frac{1}{16}'' = 1'$, you can use the milled planking sold in hobby shops for model railroad structures. It's available in a range of widths from $\frac{1}{64}$ to $\frac{1}{8}$ inch.

LAPSTRAKE HULLS

Up to now, the hull construction I've discussed has consisted of planks laid edge to edge. This method is called *carvel* planking. An important variation on the carvel-planked hull is the lapstrake hull, also known as clench nail or clinker built. This type of hull is built with overlapping strakes (Figures 6-20 and 6-21).

FIGURE 6-20. An example of lapstrake hull construction. (Photo by Patricia Leaf)

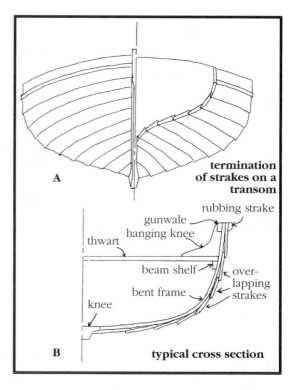

FIGURE 6-21. Lapstrake construction.

The construction of a lapstrake hull is entirely different from that of carvel-planked hulls. In this method, you plank the hull over a mold and insert bent frames after the planking is complete.

Begin by building the mold. Although you can use a solid carved mold, the built-up type is better for a planked hull; the solid type is more suitable for building up fiberglass hulls. Refer to Figures 6-22 through 6-24 during the following discussion.

Step One. Make a base from a good piece of wood or plywood with no warps. The base should be about 1 inch wider than the extreme beam of the model, and about 1½ inch longer than the model's overall length.

Step Two. Cut a spine about ¾ inch high, as thick as the keel assembly, and about ½ inch longer than the distance between the fore and aftmost section pieces. (Note that all dimensions are approximate since they depend to a great extent on the size of the model you're building.)

Step Three. Now center the spine on the base piece and glue it in place.

Step Four. At both ends of the spine, glue a piece of wood to both sides of the spine to form the fore and aft clamps. These must extend well beyond the extreme ends of the projected keel assembly.

Step Five. Make the mold sections, which are similar to bulkheads. There should be enough so that they are not more than 2 inches apart. They should be spaced more closely near the bow and stern. Extend the sections about 1 inch at the tops so as to raise the mold above the base. Notch the sections at the bottoms to accommodate the keel, and at the top to fit over the spine on the base. Put small notches (⅟₁₆ inch square) on both sides of each section at the level of the sheer line. Glue the sections upside down to the base.

Step Six. Cement thin (⅟₁₆ inch square) strips of wood into the notches on both sides of the mold sections. These strips define the sheer line (the top of the planking).

Step Seven. Build the keel-stem-sternpost assembly. The stem and sternposts are extended so that they fit into the clamps on the base (**A** in Figure 6-22). Figure 6-23 shows how to manage the sternpost of a model with a transom. Cut rabbets into the stem and sternposts, as is done for a plank-on-frame model. Figure 6-22 shows a cross-section of a typical lapstrake keel. Because of its construction, rabbets are not needed.

Step Eight. Position the keel assembly in the mold. Drill holes (⅛ inch diameter) through the clamps and stem-sternpost extensions and insert (but do not glue) a ⅛-inch-diameter dowel through each clamp (Figure 6-24). These will lock the assembly in place. When the planking is completed, you can pull the dowels out and remove the model from the mold.

PLANKING THE HULL

Now you're ready to plank. Refer to Figures 6-25 and 6-26. Earlier in this chapter I described how to use paper strips to mark the frames into equal segments. This method is ideal for lapstrake hulls because they're designed so that all strakes extend from bow to stern with no stealers.

Step One. Determine how many planks are required on each side. This should be specified on your drawing. Mark the mold sections accordingly. Starting with the garboard strake, lay a strip of tracing paper lengthwise over the mold and tick off the markings representing the strake. Refer

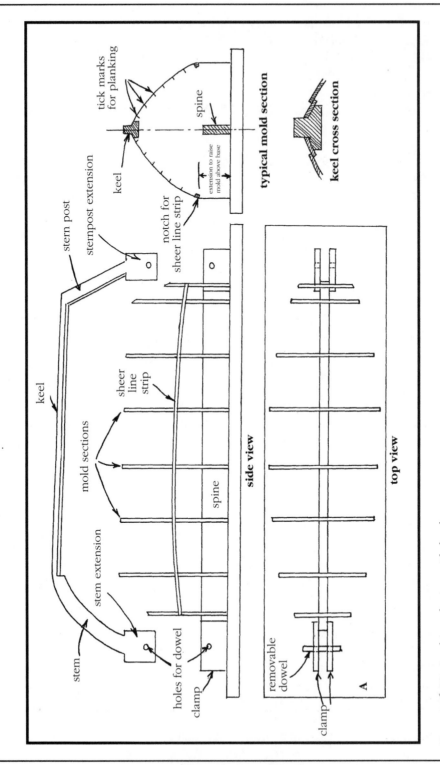

tick marks
for planking

spine

keel

typical mold section

extension to raise
mold above base

notch for
sheer line strip

keel cross section

stern post

sternpost extension

keel

mold sections

stem extension

sheer
line
strip

spine

side view

stem

holes for dowel

clamp

removable
dowel

clamp

A

top view

FIGURE 6-22. Locking device with keel.

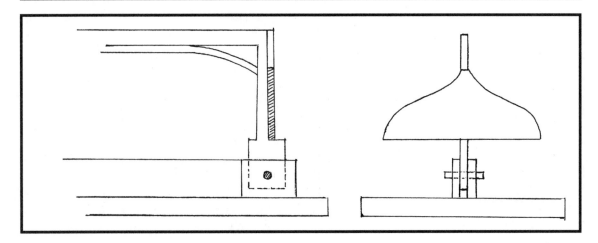

FIGURE 6-23. Locking device for a hull with a transom.

again to Figure 6-22. Connect the tick marks on the paper, and you have the shape of the strake (remember that the strakes must be a bit wider, about 1 to 1½ scale inches, than the markings indicate to allow for overlap).

Step Two. Starting with the garboard, glue the strakes to the keel assembly and to each other, not to the mold. If you want to get fancy and replicate real practice, use copper rivets to lock in the strakes. You can

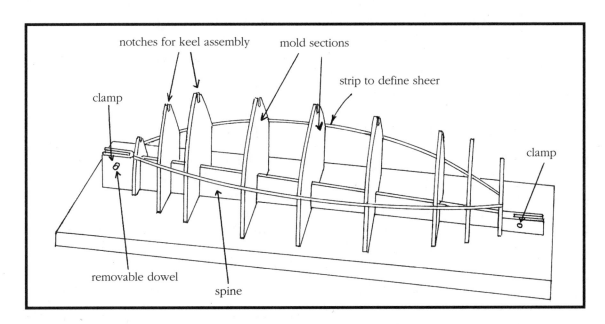

FIGURE 6-24. The completed built-up mold.

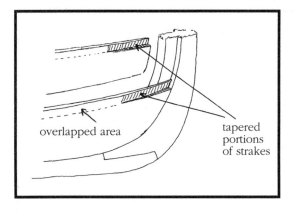

FIGURE 6-25. Forming the ends of the strakes.

make rivets from copper wire with the ends set up on the planking. As you install each strake, taper portions of its ends so that they are even at the stem and stern rabbets (Figure 6-26). Remember, if the boat has a transom, there will be no stern rabbet in the way of the transom. The ends of the strakes will be notched into the transom and into each other (**A** in Figure 6-21). Also, the edges of the strakes will be slightly beveled to provide flat landing places against each other (Figure

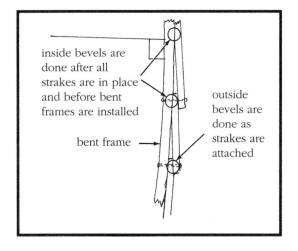

FIGURE 6-26. Beveling the strake edges.

6-26). The tapering and beveling should be done to each strake as it is installed. Put on the strakes in pairs, port and starboard.

Step Three. When all of the strakes are in place, pull the locking dowels and remove the hull from the mold. Cut off the extensions of the posts. The shaped and joined strakes will hold the form of the hull perfectly, no matter how frail the assembly may seem.

Step Four. Now you can insert the bent frames. These frames add strength only; they don't provide shaping. They should be spaced quite closely — 6 to 8 scale inches. The inside edges of the strakes should be beveled slightly so the frames will lie flat against them (Figure 6-26).

Step Five. It's now time to add thwarts and gunwales. Figure 6-27 shows a typical arrangement; your drawing will show the specifics for your model. The thwarts are transverse planks that reinforce the hull and serve as seats. Lodging knees between them maintain their spacing and reinforce them. Hanging knees provide additional support. Note in the figure that the hanging knees are notched into the gunwales. The gunwales are long planks running from stem to stern at the top of the bent frames. They help provide rigidity at the sheer line, and protect the frames. Note also the molding, or rubbing, strake on the outside of the planking. Not shown is a cap rail, which may or may not be required by your drawing. At the completion of this step, your hull is finished and ready for further detail, as required.

LIFT METHOD

If you like to carve wood, try building a lift model. This method of hull construction in-

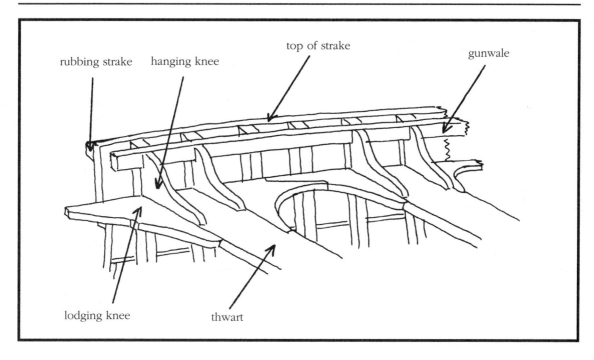

FIGURE 6-27. Thwarts and gunwales.

volves cutting slices of wood to the shapes of the waterlines, gluing them together, and then carving between the intersections of the layers, checking the sections with templates as you go (Figure 6-28).

Carve the topmost layer to the sheer and camber of the deck, and add the keel, stem, and sternpost after the hull is carved. Timberheads for bulwarks are mortised into the sides (Figure 6-29).

The more layers you use, the easier it will be to carve an accurate hull. If your drawing doesn't show enough waterlines, or if they aren't spaced correctly for the thickness of your material, develop the lines you need (see Chapter 2 for details). The drawing you use for this method must be drawn to the outside of the planking.

Hollow out the layers, except for top

and bottom, to lighten the model and to reduce stresses in the wood that might eventually cause cracking. You can do this easily by making cutouts in the layers before assembling them.

Here's a trick I use to control the use of the templates: Draw the locations of the sections on each layer with a grease pencil. When the hull is carved down, these lines appear as dots on the surface of the hull. The alignment of the dots shows the correct locations for the section templates.

PLANKS ON A CARVED HULL

Although carved hulls are an accepted method of construction in scratchbuilding,

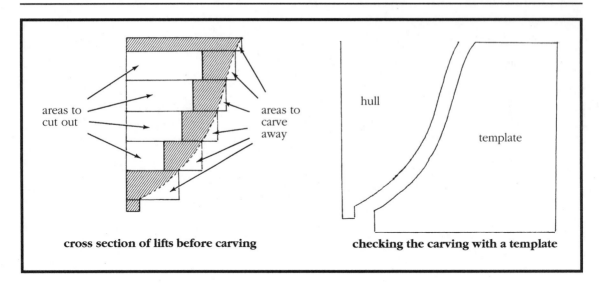

areas to
cut out

areas to
carve
away

hull

template

cross section of lifts before carving

checking the carving with a template

FIGURE 6-28. The layers of a lift model and a typical template.

they're not realistic — there's no representation of planking. One way to remedy this is to carve the hull to the inside of the planking, and then plank over it. Please don't take the shortcut of scribing the carved hull to represent planking: It violates the grain of the wood and it's hard to control. If you're prepared to go to the trouble of planking over a carved hull, however, you might just as well build a plank-on-frame hull.

FIBERGLASS HULLS

Fiberglass is a popular material for building up hulls that are intended to go in the water. In this method, you make a solid wooden mold using the lift method, and then lay up strips of fiberglass material — first diagonal to the keel and then diagonal in the other direction — until you obtain the

desired thickness. Each layer of cloth is wet out with polyester fiberglass resin until it is saturated. When all the layers are in place, use a roller to remove any uneven spots. When the material is cured, put a respirator over your nose and mouth to keep from inhaling the fiberglass particles, and sand the hull smooth. Finally, add the reinforcing frames and stringers. This method results in

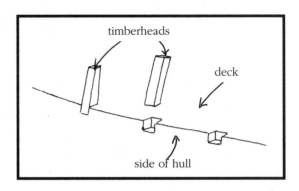

timberheads

deck

side of hull

FIGURE 6-29. Mortising timberheads.

beautiful hulls, but the technique should not be used to represent wooden hulls.

An interesting variation on the fiberglass technique is to use strips of gummed brown paper (the kind used to seal packages). Top this with several coats of Hobby Poxy inside and out, and the hull will be sufficiently waterproof. For small powered models, this method yields good results. These techniques may, of course, also be used for non-operating models, but they would not be my choice in such cases.

Vacuformed Hulls

Vacuforming a hull requires specialized equipment, and few modelers will want to get involved with this process. However, many excellent kit hulls are made this way, and you can buy ready-made hulls for a variety of models intended for radio control. The only drawback is that the models must be reinforced inside to prevent them from losing their shape.

CHAPTER 7

Miscellaneous Matters

"'Miscellaneous matters,' he calls 'em!
I calls 'em mighty important!"

— Larry Letters, local literary critic

BENDING WOOD

Regardless of which type of hull construction you choose, you'll have to bend some wood. Planks usually take the necessary curves without special treatment, but when the curves are sharp, as found at the bow and stern, the planks must be treated before bending.

Although wetting and steaming are the traditional ways of bending wood, a better method is to soak the wood for a few seconds in household ammonia. Then clamp the wood to the desired curve and let it dry.

When you release it, it'll spring back a little, but will take the proper curve easily when you glue it in place. Just don't try to glue before the plank is completely dry — the ammonia will defeat every glue. Since the ammonia alters the *structure* of the wood, instead of merely softening the wood, you can obtain very tight bends with this method.

COPPERING

If the ship you're modeling was coppered below the waterline, you should copper the model to achieve authenticity. Some writers have said that coppering is impractical at less than ¼"=1' scale. Nonsense. You can copper beautifully at ⅛"=1' and even smaller.

Step One. The standard size of copper plates for real ships was 18 inches wide by 4 feet long. On a 0.0001- to 0.003-inch-thick sheet of copper, lightly score the outlines (to scale) of your plates in the form of a grid. This makes them easy to cut out.

Step Two. Cut out the plates with a sharp knife and metal ruler. (You can use scissors or a paper cutter, but these might not give smooth edges.) Bear in mind that you are cutting individual plates. I don't agree with writers who recommend that you fold long strips of copper to simulate overlaps, or that you scribe the strips to simulate the separations between plates. Folding produces lumps that are out of scale, and scribing doesn't produce the needed fore and aft overlaps. Long strips are also difficult to fit properly.

Step Three. Treat the plates to eliminate the shine of new copper. There are several ways to do this; choose the one that yields the results you're after:

- To achieve the mottled blues, greens, and purples evident in copper that is not quite new, bake the plates in a 450-degree oven for 30 minutes before installing them.
- To achieve the greenish patina of old copper, paint the plates *after they have been installed* with a super-saturated solution of salt and vinegar. Let it sit overnight, wipe off the film, and repeat the process until the desired effect is obtained. Seal with a coat of polyurethane varnish.
- For aging copper, you might want to investigate commercial compounds, which yield different finishes.

Step Four. Decide whether to show nail heads. Many modelers try to simulate the heads of the nails used to fasten the copper plates on real ships. They do this by using a toothed wheel to press rows of raised dimples around the edges of each plate. But the fact is, the nail heads on real ships were not raised; they were driven home to form a slight, almost invisible, *concavity* in the copper. These might show up on a large-scale model, but it's not worth doing at less than a ⅜"=1' scale. Many fine modelers will disagree with me, but I stand firmly on the side of authenticity.

Step Five. Now you're ready to lay out the pattern of the copper plates on the hull. The top of the copper is usually about one foot above the load waterline, so the first step is to establish the top of the copper, even though you'll actually be starting at the bottom when you apply the plates.

To mark the waterline (or the top of the copper), you'll need a working cradle to hold the hull steady and in the correct position (mentioned in Chapter 6), and a means of holding a soft pencil (HB is best) at a constant height. Gluing or taping the pencil to a block of wood does the job, or you can make or buy a simple adjustable jig. Both options are shown in Figure 7-1. I recommend that you make the jig shown in the figure (my own design), since it can be used every time you build a model.

Step Six. Establish the pattern for the coppering, which is influenced by the shape of the hull. The courses of the copper *do not* follow the lines of the planking (Figures 7-2 and 7-3 show typical patterns; note that the letters on Figure 7-2 correspond to the ones on Figure 7-3). You don't have to lay out every row, just the general flow and the gore lines (the lines at which the pattern of

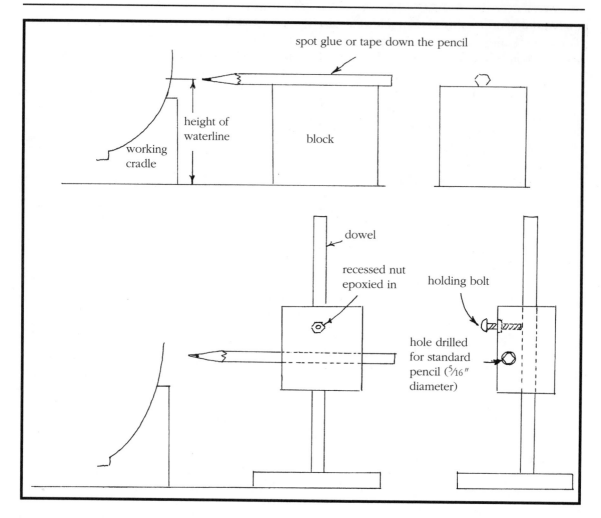

spot glue or tape down the pencil

height of
waterline

block

working
cradle

dowel

recessed nut
epoxied in

holding bolt

hole drilled
for standard
pencil ($\frac{5}{16}''$
diameter)

FIGURE 7-1. Methods for marking the waterline.

the courses changes). Because of the curvature of the hull, successive parallel courses of plates would result in the ends of the courses curving up radically toward the ends. This would not look good, and would cause difficulties in meeting the top of the copper neatly.

Step Seven. Since the upper plates overlap the lower ones, and the forward plates overlap the after plates, start attaching the plates to the hull at the keel and sternpost (Figures 7-4 and 7-5). Use epoxy, Ambroid, or contact cement. Note also that the first row on each side of the keel slightly overlaps the keel or sternpost.

Step Eight. When the hull is com-

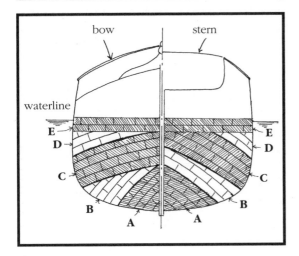

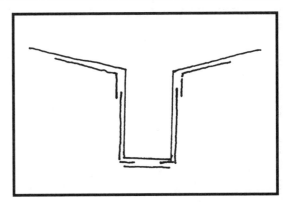

Figure 7-4. Coppering the keel.

FIGURE 7-2. Coppering patterns (bow and stern views).

pletely coppered up to the bow, plate the cutwater as you did the keel so that the plates overlap the plates behind them.

STEEL PLATING

Modelers go to great lengths to show the planking and coppering on a wooden ves-sel, but when it comes to a riveted steel ship, they're often content to leave the hull smooth. Strange, because the processes are very similar. On a scale of $\frac{1}{8}''=1'$ or larger, you can use index card stock or a thin sheet of copper or aluminum. Unlike coppering, however, the rivet heads do show on real ships with steel plating. You can show this on your model by indenting the plates from the back. Use the same method for deter-mining the courses of the steel plates as you did for the copper plates, but note the dif-ference in the overlap pattern (Figure 7-6).

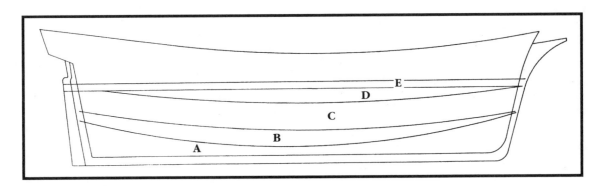

FIGURE 7-3. Coppering patterns (elevation view).

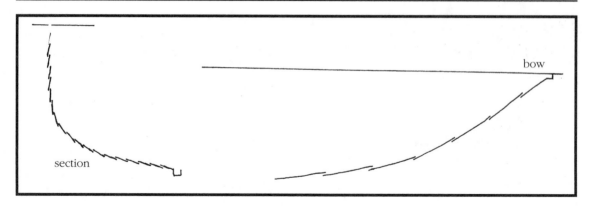

Figure 7-5. Overlapping the plates.

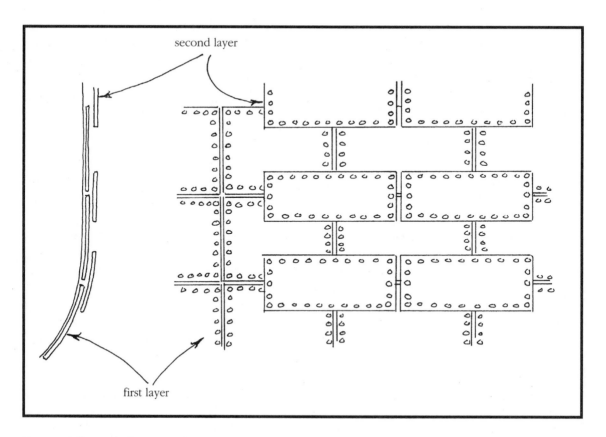

FIGURE 7-6. Steel plating.

Propeller Shaft Fairings

Propeller shaft fairings on ships with more than one propeller pose a problem. Carving them to shape and putting them on after the hull is planked creates a nightmare trying to position and align them properly. Here's a good way to do it.

Step One. Look on the section drawing for the cross sections of the shaft fairings.

Step Two. When you make the bulkheads, be sure to include the fairing cross sections, as shown in Figure 7-7.

Step Three. Cut the fairing portion of the bulkhead to the *outside* of the planking, and cut the rest of the bulkhead to the *inside* of the planking as usual.

Step Four. Install the bulkheads and plank the hull, except for the fairings, as shown in Figure 7-7.

Step Five. Now fill in the spaces between the fairing sections with blocks of wood, and carve to shape.

Locating Propeller Shafts

If you have multiple propellers for which shafts exit the hull without significant fairings, you can adapt the preceding method to locate the shafts.

Step One. Using the section drawing, locate the points on at least two bulkheads through which the lines of the shafts will pass (Figure 7-8).

Step Two. Before installing the bulkheads, drill them to accept brass tubes whose inside diameter will accommodate the propeller shafts.

Step Three. After the bulkheads are installed, insert the brass tubes and cement them in place. CA glue is good for this.

Step Four. Plank the hull. When you come to those areas where the shafts exit the hull, temporarily insert the shafts to determine where you'll have to drill holes in the planking.

Step Five. After the hull is completely planked, sanded, and primed, insert the shafts permanently with CA glue.

Bilge Keels

The bilge keels found on some modern ships are a nasty problem. Viewed from the side they look as if they curve upward at the ends (**A** in Figure 7-9). If you look fore and aft, however, you'll see that they are actually straight (**B** in Figure 7-9).

The bilge keels lie along the ends of a diagonal. Therefore, to determine their shape, you need to plot the end points of the diagonal involved (described in Chapter 2 under the heading "Checking for Accuracy"). This plot will be the inside curve of the bilge keel. The outside edge will be parallel to the inside edge.

You must be very careful to ensure a good fit against the hull. Plot the ends of the appropriate diagonal on the bulkheads before installing them. When planking, when you get to the point where the planking reaches these points, spot them on the surface of the planks and use them to align the keels, which are cemented to the outside of the planking. You might have to make some final adjustments to the inside edges of the keels, so I suggest you leave extra material, and finish the outer edges only after the inner edges fit well.

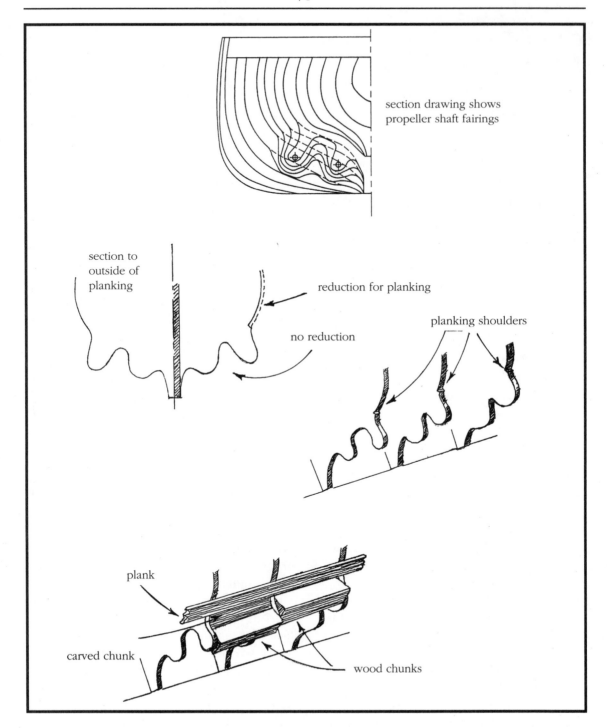

section drawing shows propeller shaft fairings

section to outside of planking

reduction for planking

no reduction

planking shoulders

plank

carved chunk

wood chunks

FIGURE 7-7. Propeller shaft fairing section.

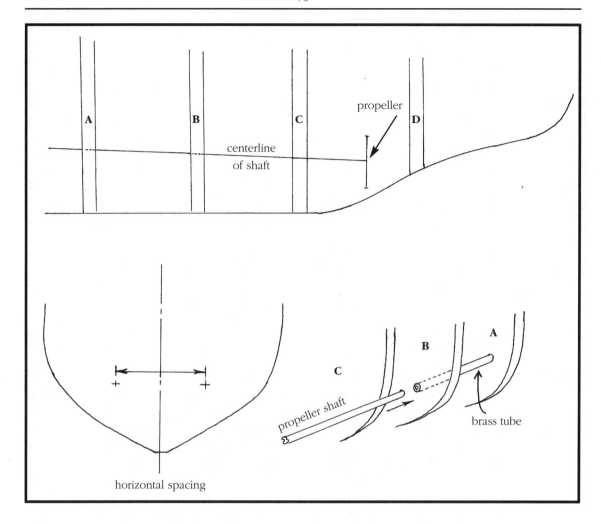

FIGURE 7-8. Locating propeller shafts.

DECK FURNISHINGS AND HULL FITTINGS

When you've finished the hull, you're ready to do the detail work that can make or break a model. I've seen many a nice hull spoiled by indifferent, or just plain crude, detail work.

To the deck you can add furnishings (deck houses, hatch coamings, knightheads, and binnacles) and fittings (capstans or windlasses, bitts, wheels, and pumps). The outside of the hull gets a rudder, channels for the shrouds, catheads, trailboards, and figureheads, to name a few of the many items that make up the detail.

Although experienced scratchbuilders

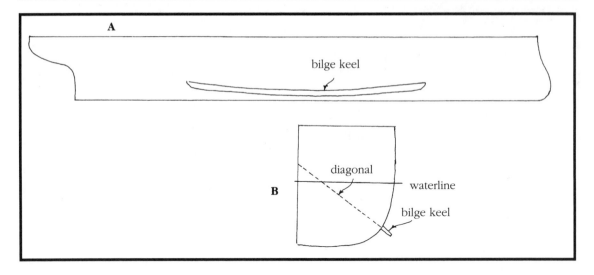

FIGURE 7-9. Bilge keels.

will want to make many of these items by hand, those new to scratchbuilding will probably rely more heavily on commercial fittings. When properly scaled and appropriate to the specific model, commercial fittings are perfectly acceptable. Ship's wheels, for example, are difficult to make and good ones are readily available. Capstans, riding lights, cowl ventilators, grating, and other hard-to-make fittings are also available. If you're intent on making as many of your own furnishings and fittings as possible, refer to the list of books contained in Appendix II; many provide excellent guidance in detail work.

If you're making your own furnishings, take your time. Make clean cuts, and finish the wood carefully. Here are a few do's and don'ts: *Don't* gum up the detail with heavy paint (more on this in Chapter 8), *don't* glue on bits of paper with windows and doors drawn on them, and *don't* "represent" paneling with ink lines. *Do* build up paneling

with thin wood or card stock. Take care to fit corners carefully, and if they should be square, be sure that they are. Be accurate when the bottom edge of a structure must be curved slightly to fit the camber of the deck. And *do* ensure that everything is properly scaled.

ATTACHING THE RUDDER

The best way to attach the rudder is to make working gudgeons and pintles rather than attempting to represent them. You can make very small ones using the following instructions (refer to Figure 7-10).

Step One. Using the cutoff disk on your rotary power tool, slice off pieces of brass tubing as long as the gudgeon and pintle straps are wide. (The diameter of the tubing and the thickness of the straps depend on the scale of the model.) You'll need a piece of tubing for each gudgeon and pintle.

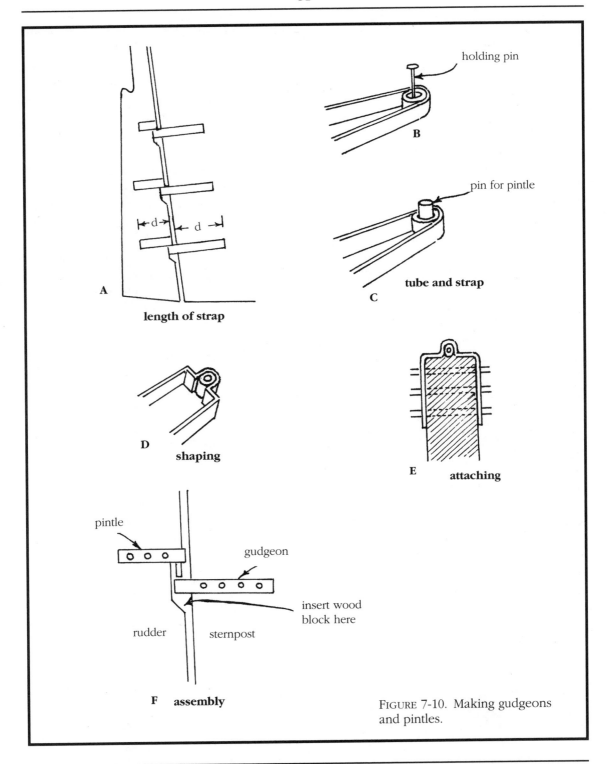

FIGURE 7-10. Making gudgeons and pintles.

Step Two. Cut brass straps about three times the dimension "d" in **A** of Figure 7-10.

Step Three. Bend the straps around lengths of tube (**B** in Figure 7-10). Pin the tubes in place while you work with them.

Step Four. With a soldering iron, tin the inside of the straps at the bend and one side of the tube. Solder them together.

Step Five. If you're making pintles, cut pieces of brass wire twice the length of the tubes and solder them into the tubes (**C** in Figure 7-10). Gudgeons have hollow tubes.

Step Six. Now you can shape the straps to the rudder or sternpost (**D** in Figure 7-10). Trim the straps to the correct length and then attach them:

- If the model is small scale, press dimples into the straps to represent rivet heads, and glue them on with epoxy or CA glue.

- If the model is large scale, drill the straps for tiny brads, and nail them on. If brads appear to be out of scale, drill through the rudder/sternpost, insert pieces of brass wire all the way through the straps on both sides, solder each end, and file the ends down (**E** in Figure 7-10).

Step Seven. If you've positioned the gudgeons and pintles correctly, attaching the rudder is easy. (Remember that the pintles are on the rudder, the gudgeons on the sternpost.) The length of the cutout on the rudder is about three times the width of the straps. Simply position the pintles over the gudgeons, lower them into place, and cement a small wood block beneath the gudgeon to prevent the rudder from slipping off (**F** in Figure 7-10).

CHAPTER 8

Applying the Finish

*"Well, make up your mind! Yellow, green,
red, puce . . . why don't we paint it black?
Black is a great color!"*

— Dan Dauber, ship painter

The first step in applying a finish to a model is to decide whether to use paint or natural wood finishes. The choice, of course, is dictated to a degree by the wood you used to build the model. Holly and pine, for example, are normally unsuitable for the natural treatment. Apple, boxwood, and walnut, on the other hand, are striking when treated to a natural finish.

Personally, I prefer paint. Some modelers maintain that paint is merely a convenient coverup for faulty workmanship. This might be true sometimes, but certainly not always. In most cases, a natural finish doesn't reflect the appearance of the real ship and thus violates the integrity of the model. Paint is more realistic, and the contrast of colors picks up details and enhances appreciation of the model. The choice, however, depends on your personal taste, so you'll have to weigh the arguments and make up your own mind.

Whatever finish you choose, be sure to apply it as you go. You can't build the entire model and then go back and paint it. Do the hull and decks as soon as the structure is complete. Assemblies such as deckhouses and hatches are done before they are installed. Spars are done before their hardware is installed, and before they are put on the model. Use common sense — always paint when you have the easiest access to the area.

MATERIALS

In addition to the tools mentioned in Chapter 4, you'll need some specialized tools just for applying the finish. Keep in mind that even though you decide to paint the overall model, there still might be a few items you want to stain.

BRUSHES AND RAGS

The importance of good brushes can't be overemphasized. Though expensive, sable brushes last long enough to justify their price (if you take care of them). Cheap brushes are no bargain: they fall apart easily and give poor results. You'll also need some soft absorbent rags for applying stains. Some people use brushes for the stain, but I find that rags do a better job.

MASKING TAPE

Masking tape, though essential for such jobs as painting waterlines, is too wide and too coarse for model work. It's hard to make it conform to the curves of a hull, and the thickness causes paint to build up along the edge in a ridge that you have to later sand down. Finer tapes are available, but they're not easy to find. The best I've found is a narrow plastic tape used for pinstriping. It stretches and conforms well to curved surfaces.

If you must use ordinary masking tape, reduce the thickness of the edge: Stretch a length of tape on a cutting board and, using a straightedge, cut the tape in half lengthwise with a sharp knife held at an acute angle to the cutting surface. By putting a taper on the edge of the tape in this way, you can virtually eliminate buildup. You should also reduce the width of the tape to about ¼ inch so that it follows the curves of the hull more easily.

PAINTS

For best results, choose paints made for modeling: Humbrol and Floquil model railroad paints are outstanding, Testor's Model Master and Pactra are also fine, and even the water-soluble acrylics sold in hobby shops are acceptable.

Artist's acrylics, sold in tubes in art supply stores, are the best for painting such things as carvings and flags. These colors are very thick as they come out of the tube and have to be thinned with water to a workable consistency. You can thin them down to the consistency of heavy cream for opaque painting, or to a watery consistency for transparent wash work (to age or weather the model). You can also mix them easily to get exactly the right color. If you use acrylics over plastic or metal, be sure to prime first with a metal primer or a good flat enamel.

One last note on paints: Use matte colors almost exclusively. Glossy colors are inappropriate on most ships. Semi-gloss is acceptable for some applications on modern yachts.

STAINS

In spite of the variety of commercial stains available, I prefer to make my own. You can achieve exactly the hue and tone you want, and you can make as little or as much as you need so you won't have a lot of partially used cans of stain lying around. To make the stains, use artists' oil colors. Mix the colors until you get the hue you want, and then thin the mixture with turpentine. Vary the tone (intensity) of the color by thinning more or less. Don't be afraid to experiment. Test the colors on scrap wood until you see the results you're after. The most useful colors for making ship-model stains are burnt umber, burnt sienna, yellow ochre, and ivory black. Raw ochre, cadmium red (medium), and cadmium yellow (medium) are also useful.

THINNERS

Always keep a stock of thinners on hand; you'll use them to both thin the paint and clean your brushes. Be sure, though, that the thinners are compatible with the brands of paints you use. Floquil paints, for example, use a different base from those used in other paints; using Floquil thinner on Testor's or Pactra paints results in a gummy mess.

PRIMERS

For wood, the best and least expensive primer is plain flat white paint. For metal surfaces, you need a primer specially made for these surfaces. Check your local hobby shop.

VARNISH

In most cases, matte or satin finish polyurethane varnish is best. On modern yachts, where a high finish is desired, use gloss varnish and then rub it down with a soft, lint-free cloth.

OIL FINISHES

Natural and stained wood hulls and decks can be oiled instead of varnished. Linseed and olive oils have been recommended for years. An excellent and traditional method is to use a 50-50 mixture of steam-distilled turpentine and boiled linseed oil into which small pieces of beeswax have been dissolved. This is probably what is referred to in old books as *French* polish, and was used on the British Admiralty models. Tung oil, sold in paint stores, is also highly recommended.

HOBBY POXY

Hobby Poxy is a trade name for a two-part epoxy paint. It comes in many colors, though most are unsuitable for ship modeling. The clear Hobby Poxy, however, is excellent for waterproofing. If you're building a working model, apply a few coats inside the hull and over the outer paint and you can safely put your model in the water.

TECHNIQUES FOR APPLYING THE FINISH

PAINTING

If you decide to paint your model, you'll need to prime the hull, mark the waterline, and mask it before painting. The technique for marking the waterline is described in Chapter 7 under the heading "Coppering."

The secret to a good paint job begins with a properly prepared surface. Begin by sanding the surface smooth, cleaning the wood with a tack cloth to remove all dust, and then applying one or more coats of primer, sanding and tacking between each coat. This preparation is vital to a good paint job; I cannot overemphasize this point.

Now you're ready to paint. Apply the first *thin* coat of paint. Don't try to save time by applying the paint thickly — several thin coats are infinitely better than one thick coat. Stroke the brush in one direction only, and brush *into* already painted areas, not *away* from them. Sand this coat thoroughly with dry fine-grit paper (don't worry if you sand through to the wood in places).

Apply the second thin coat of paint. Sand it thoroughly with *wet* sandpaper — after the first coat, always use wet sandpaper in increasingly finer grit. Repeat this process until you you're satisfied with the results. After the final coat, I recommend using fine pumice powder instead of sandpaper: Dampen a soft rag, touch it to a bit of the powder, and rub the surface gently until you have achieved a uniform finish. If you can't find pumice powder, 400- to 600-grit wet sandpaper will do. Finally, wipe the surface with a clean damp rag to remove any residue.

I don't recommend using steel wool for model work. The fibers get into everything, and it's a mess to clean up. You might want to try bronze wool, which is available in boating supply stores — it does the same thing as steel wool, but more gently, and isn't as messy.

The painting process is involved, to be sure, but certainly worth the effort for large surfaces such as hulls and rails. Small pieces, such as hatches and deckhouses, don't require so much work. Often one primer coat and one paint coat are enough. Any more coats on these small pieces and you risk obscuring detail with paint buildup.

STAINING

Stains are applied directly to the bare, dust-free wood with either a soft brush or cloth. Avoid getting stain on areas to be painted, such as a stained mast that is to have a painted white masthead. Stain is hard to cover with white paint — it tends to bleed through. If you wish, apply varnish when the stain is dry. The varnish seals the wood and relieves the dull appearance of some stains.

VARNISHING

Varnishing is no longer the tedious and exacting chore that it used to be, thanks to the new fine-grained polyurethane varnishes. Use a tack cloth to ensure a clean surface. *Stir* the varnish carefully (shaking the can forms bubbles, which could ruin your work). Flow the varnish onto the clean wood with a soft brush, working in one direction only and brushing into previously covered areas. When dry, sand the first coat smooth with *wet* fine-grit sandpaper, clean the surface again with the tack cloth, and apply the second coat. When dry, rub it down once more with very-fine-grit wet sandpaper. Two coats of varnish is sufficient.

OILING

Oiling can be done over either stained or bare wood. An oiled finish is highly recommended over fine woods. Moisten a soft cloth with the oil and rub it into the wood. The first coat will probably raise tiny wood fibers, giving a slightly rough texture. Sand gently with wet fine-grit sandpaper, and apply another coat. When dry, burnish with a clean, soft, lint-free cloth.

WEATHERING

Weathering is the art of painting a model so that it looks aged. Model railroaders are masters of weathering techniques because they strive for realism in their layouts. Most ship modelers never attempt weathering, which is a shame because it can add real character to a model. I once did a model of a tramp steamer and added extensive weathering; it attracts favorable attention from everyone who sees it.

Use acrylic paints for weathering. Thin gray washes look like salt residue. Slightly thicker mixtures of burnt sienna and cadmium red simulate rust. Thin, stippled black around a coaling hatch looks like coal dust. A thin mixture of yellow ochre and viridian looks like mold on wood. The possibilities are endless. In applying weathering, look for places where salt spray and rust might accumulate, such as under scuppers and hawse holes, around the bow, and in corners. If something rusts high up, there's usually a rust stain running down from it. White surfaces gray irregularly, as do black surfaces. There should be no pure, bright colors. The sides of the ship might be marred from rubbing against docks. In the case of iron or steel ships, these areas might show rust. Thin areas of black appear around the winches where oil has dripped. On bitts and winches, the paint might have worn off some metal surfaces, exposing the metal beneath. These are only a very few suggestions. If possible, look at real ships closely. Use your imagination, but temper it with common sense.

AIR BRUSHING

Air brushing is a superior way to apply paint, especially if you want a smooth, uniform surface. It is, however, a skill unto itself. If you intend to attempt air brushing, buy good equipment or you'll quickly become discouraged. Get a book on airbrushing techniques, and arrange to see a demonstration. Then practice before you try it on a model.

GOLD LEAFING

Because of the perceived high cost and the "mystery" of the gold-leaf process, many modelers are content to use gilt paints instead. But consider this: A book of gold leaf costs about $50, and contains so much that you'll probably will it to your grandchildren. The process is not complicated, either — it simply requires care and patience like everything else.

Step One. Clean the surface carefully with acetone and brush on a thin layer of adhesive made for the purpose. This adhesive, which is available in good art supply stores, is clear; professionals mix it with a bit of yellow color (also made for the purpose) so that they can see which areas have been coated.

Step Two. Let the adhesive dry for several hours until it's just a bit drier than tacky.

Step Three. Gold leaf is so thin and delicate that it can only be handled while on its paper backing. If you touch the leaf itself, it's apt to crumble into dust. To apply it, tear off a small piece (for most model purposes, about ¼ inch to ⅜ inch square) with tweezers, lay it face down against the surface, and then remove the backing. Repeat the process until the area to be gold-leafed is completely covered. Small overlaps are permissible. If you're leafing a complex area, such as the face of a figurehead, you might need to use smaller pieces so that the gold leaf conforms to the contours of the surface.

Step Four. Finally, with a good small brush, buff the gold leaf well onto the surface. The appearance is vastly superior to gilt paint, and gold leaf won't discolor with time.

A caveat: In the old days, gold leaf was rarely wasted on small vessels, unless they were royal yachts or the like; yellow paint was frequently substituted. Keep this in mind if you want an authentic appearance for a 10-gun schooner that appears to have gold decoration.

LETTERING

Lettering on a ship model is usually confined to the name and the port of registry on the stern, and sometimes the name on the bow. This doesn't sound like much, yet it can be one of the most demanding jobs in finishing a model. For one thing, there are no hard-and-fast rules on lettering style. You'll have to look at photos, drawings, or paintings of contemporary vessels and make your own decisions.

If the model is of a modern ship, you can often use press-on letters, which are available in many styles and sizes in art supply stores. Choose an appropriate style and make sure the letters are correctly aligned on the model.

Press-on letters are great, but for a model of an older ship, you might not be able to find an appropriate style. You'll probably have to paint the letters yourself. To do this, first lay a piece of thin paper on the surface to which the lettering will be applied, and sketch out the top and bottom lines for each row of lettering. Since most transoms have a curve, these will not be straight lines. Do the lettering on the paper, transfer it to the model, and then paint the letters.

I recommend acrylics and a high-quality small round sable brush for lettering. Great care is required, and you'll probably have to do some touch-up. You can make this process easier by painting very thin paper with the background color of the model, painting the lettering on this paper, and then cementing the paper to the model with carpenters' glue. Use *dry* fine-grit sandpaper to feather the edges of the paper so that it blends in with the hull surface. Touch up the paint when done.

The most demanding task arises when you know that the lettering was carved in relief. This is something that only the most accomplished miniature woodcarvers should attempt. If you try it, use boxwood or pear, very sharp miniature woodcarving tools, and a great deal of patience. If your hull is made of a soft wood, you can insert pieces of boxwood or pear where the carving is to be done.

CHAPTER 9

Masting and Rigging

"This here splice was made by a sojer with ten thumbs and a twisted marlinspike!"

— Benny Bowline, frustrated boatswain

Don't be discouraged if the only masting and rigging information immediately available is a sail plan, a schematic rigging plan, or merely a table of spar dimensions. You can fill out even the scantiest information with a little research. Appendix II lists many books that'll help; some give complete proportions for masts and spars, others show what hardware is required and how the various lines should be handled. Don't necessarily assume that the information is always accurate, though. Actual practice varied considerably according to time and place. Don't introduce anachronisms. Be sure of the dating of any reference you use.

Spars

The first rule in making spars from scratch is not to use dowels. Dowels are often made from poorly grained wood that's hard to work and apt to warp and split. Furthermore, the shapes of spars are complex, and it's easier to control the shapes if you can plot them on a block of wood.

Start with a piece of straight-grained wood whose dimensions are a bit greater than the outside dimensions of the spar you're going to make. Trace the profile on one side. Then cut it out and sand it to shape (I do most of the shaping on a sanding machine). Rotate the piece 90 degrees and repeat the process. Now carve and sand the piece to the correct cross-sections. Read your drawing carefully — there might be transitions along the length of the mast from square to octagonal to round.

Except for small craft, ship's masts are

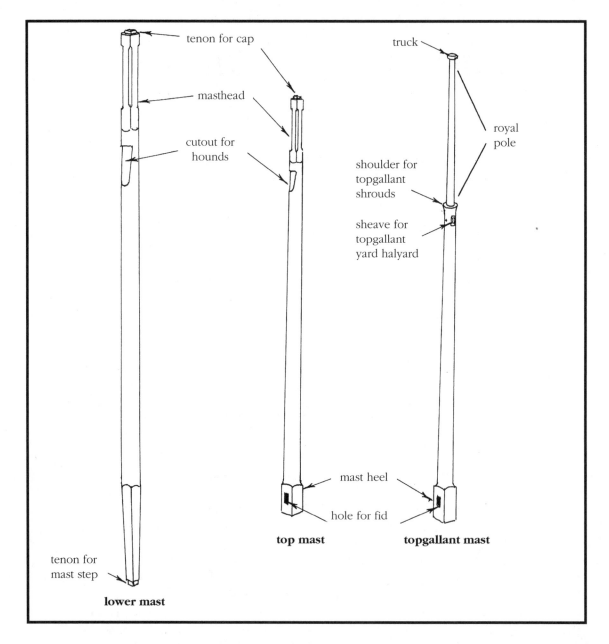

FIGURE 9-1. Typical masts.

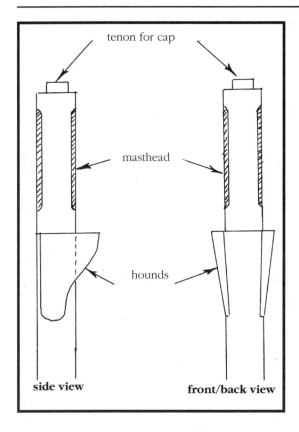

tenon for cap

masthead

hounds

side view   **front/back view**

FIGURE 9-2. Masthead with hounds.

made in sections: a lower mast, a top mast, often a topgallant mast, and sometimes a royal (Figure 9-1). The topgallant and royal are frequently the same pole, on which both the topgallant and royal yards are hoisted. The lower masts on large ships were frequently built up of wedge-shaped sections because of the difficulty in finding large enough tree trunks. Such built-up masts had rope lashings or iron bands around them to hold the assembly together. If your ship had such bands, be sure to include them.

When making masts, except for the top one, you'll have to cut the surfaces and ledges on which the hounds rest, shape the head, and cut a tenon in the top of the mast-

head for the cap (Figure 9-2). Be careful to fit and shape the hounds correctly. Next build the crosstree, the trestletree, and the top platforms, if any. At the top of the topmost mast is a *truck* — a piece of wood that protects the end grain of the mast.

The bowsprit, strictly speaking, is the spar that protrudes directly from the bow of the ship. It's extended by a jib boom, and sometimes a flying jib boom. You assemble these booms using caps and tenons in much the same way you assembled the masts. Figure 9-3 shows a typical bowsprit/jib boom/flying jib boom assembly.

Notice the martingale and spritsail yard in the figure. The martingale supports the stays that hold the jib boom down, and the spritsail yard spreads the jib boom guys. In older ships the martingale (also known as the dolphin striker) is double, in the shape of an inverted V. The spritsail yard, which originally supported a sail, was replaced in the nineteenth century by the whisker boom — a shorter, less bulky spar — though it was not always used.

The bees behind the bowsprit cap hold the forestay in place. The flying jib boom is held by an iron fitting rather than by a cap (though this was not always the case), and is mounted to one side of the jib boom so that it doesn't interfere with stays that pass through the jib boom.

In assembling masts and bowsprits, you have to make caps. This can be difficult because of the relatively large holes you need to make in small pieces. To keep the wood from splitting, first laminate two pieces of wood, cross-grained and larger than the cap (**A** in Figure 9-4), to make up the thickness of the cap. Lay out the cap, drill the holes, and then trim the wood down to the outer shape of the cap (**B** in the figure).

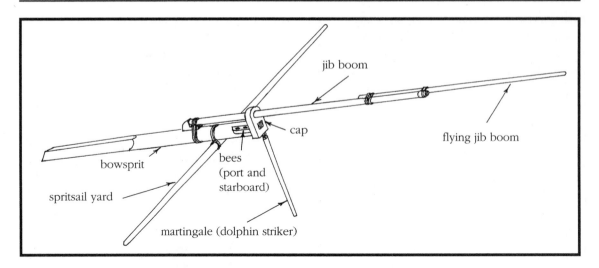

FIGURE 9-3. A typical bowsprit assembly.

ASSEMBLING MASTS AND SPARS

Figure 9-5 shows a typical assembly of the lower and upper mast. Note the fid (which prevents the upper mast from sliding down), and the bolster (which provides a smooth rounded surface for the shrouds to ride

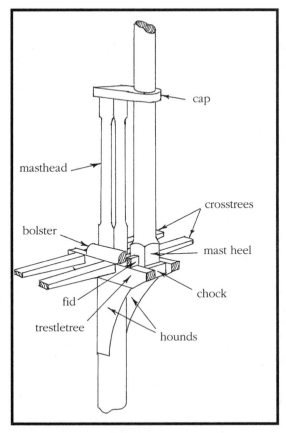

FIGURE 9-5. Assembling the upper and lower masts.

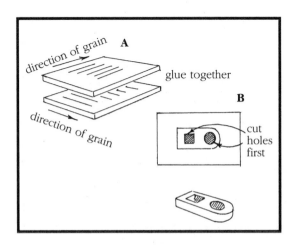

FIGURE 9-4. Making a mast cap.

over). The exact shape of the cap varied according to nationality and time. When iron fittings became common in the nineteenth century, the wooden cap was replaced by an iron device.

Use the same type of wood for the yards as you did for the masts. The center section of larger yards was frequently octagonal in cross-section, transitioning to round. Smaller yards often had wooden yokes fastened to the back of the yard. These rode against the mast, and the parrals, which secured them to the mast, were attached to them. Yards usually had thumbcleats near the ends of yards to prevent the movement of rigging. All yards had hardware of some sort, which became more complex as iron fittings became more common. Figure 9-6 shows typical yards and their hardware. Notice the jack-stay, a relatively recent innovation. The heads of the sails were attached to the jack-stays rather than directly to the yards, as was the earlier practice.

If you're making gaffs or booms, there are jaws to consider. There are several basic patterns for jaws, and gaff jaws are somewhat complex. For one thing, gaff jaws can be bent. This was done so that the jaw sits at a more or less right angle to the mast when the gaff is hoisted. Make them carefully since they tend to be delicate. Make the parrals and attach them to the jaws (Figure 9-7). These will bind the jaws to the masts. Gaffs and booms also had thumbcleats to prevent the movement of lines attached to them.

Be sure to assemble the masts and other spars, and to attach any hardware required, before installing them on the ship.

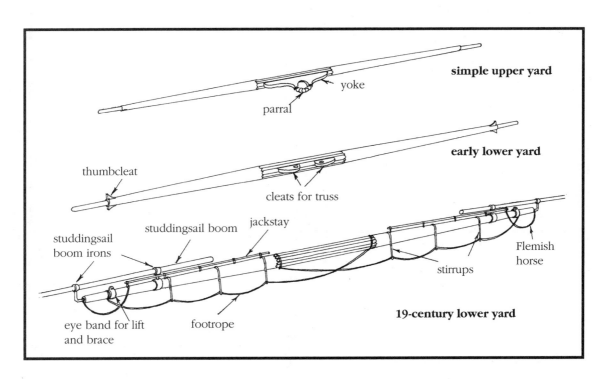

FIGURE 9-6. Typical yards.

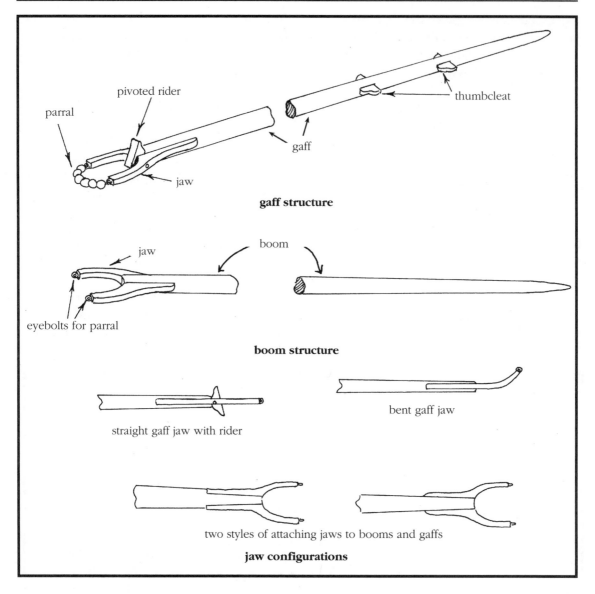

gaff structure

boom structure

straight gaff jaw with rider

bent gaff jaw

two styles of attaching jaws to booms and gaffs

jaw configurations

FIGURE 9-7. Typical boom-and-gaff constructions.

It's not easy to do these things once they're aboard. If you've placed the mast steps and partners correctly, you'll have no difficulty installing the masts so that they stand at the proper angles. It's not necessary to install the yards at right angles to the ship's center-line; you can brace them around at an angle instead. This adds interest to the ship, and also reduces its overall width — often an advantage in displaying the model.

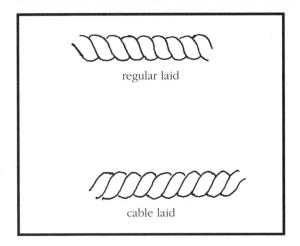

FIGURE 9-8. Cable-laid and regular-laid line.

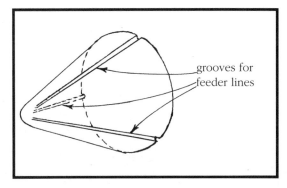

FIGURE 9-9. A rope-making top.

ROPE AND ROPE MAKING

Rigging a ship requires rope in a wide range of sizes in both black for standing rigging and brown for running rigging. Please don't put white line on a model; rope is yellowish-brown. Fabric stores sell thread in a color called "camel," which is perfect.

There's also the consideration of cable-laid and regular-laid line (Figure 9-8). Cable-laid line has a left-hand twist; in naval service cable-laid lines were used for shrouds and stays, and in merchant service they were used for stays only. Regular-laid line has a right-hand twist; it was used for virtually everything else. Many modelers ignore the distinction, and on small-scale models the difference is hardly noticeable. It's a nice touch of authenticity, nevertheless.

If you're after realism in your model, or if you cannot otherwise find the correct sizes, you might want to make some of your own rope. As mentioned in Chapter 4, rope-making machines are not widely available, so you might have to make your own machine. The basic principles of such a device

are fairly straightforward: The rope is twisted up from three lines known as *yarns*. The yarns are not simply twisted around each other, though. They are twisted individually before being twisted together. The feed of the yarns is controlled by a simple device called a *top* (Figure 9-9).

At one end of the machine is the headstock, which consists of three gear-driven hooks to which one end of each yarn is attached. (The gears can be cranked by hand or motor driven.) The headstock twists the three yarns independently and simultaneously (Figure 9-10).

At the opposite end is the tailstock (Figure 9-11). Here we have two tensioning devices: One, attached to the top, maintains the pressure of the top against the yarns as they are twisted and made into rope; the other, equipped with a swiveling hookup, keeps the yarns taut.

The top does not turn. Figure 9-10 shows one way to control the top automatically. Notice the arms on the top, which ride on a pair of tightly stretched cords or wires. By putting a handle on its back, the top can also be hand-held, and thus manually controlled, on both large motor-driven

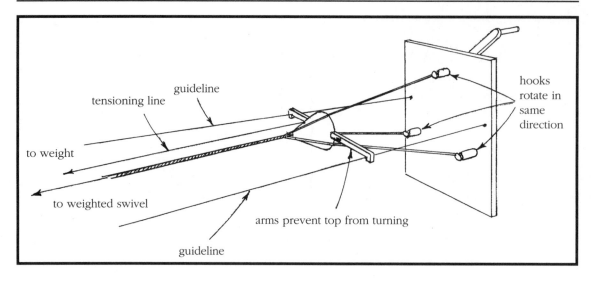

FIGURE 9-10. The headstock of a rope-making machine.

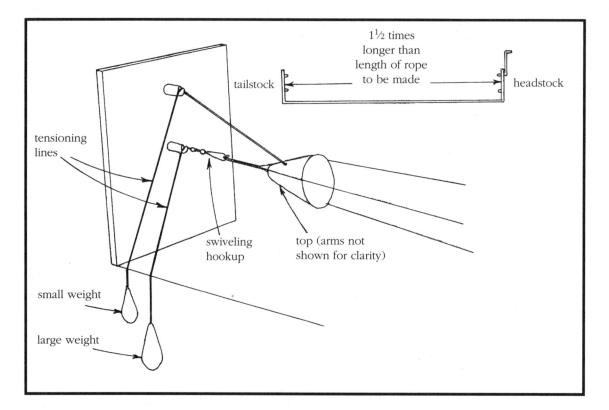

FIGURE 9-11. The tailstock of a rope-making machine.

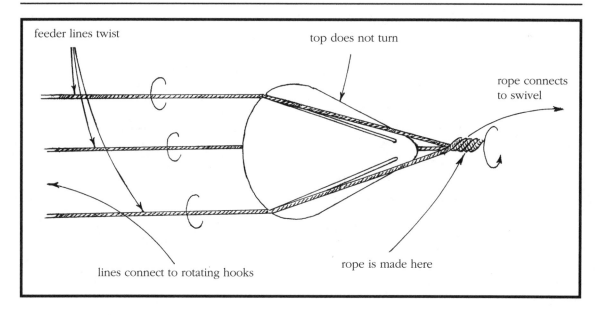

feeder lines twist

top does not turn

rope connects
to swivel

lines connect to rotating hooks

rope is made here

FIGURE 9-12. Making rope.

machines and on hand-cranked machines. On a hand-cranked machine, however, the operation then requires two people: one to crank and the other to control the top.

When you begin twisting the yarns, nothing seems to happen at first. As the tension is increased on the top by this twisting action, however, the yarns begin to twist around each other at the swivel, and the rope begins to form (Figure 9-12).

Now that you know how a rope-making machine works, with a little ingenuity, you'll probably be able to devise your own machine. It can be simple or elaborate, but simple usually works best. Keep in mind a few important points:

- The distance between the headstock and the tailstock must be about 1½ times the length of the rope you want to make because of the shortening that occurs when the rope is twisted.

- The tension on the top and on the rope is critical. If you use weights, experiment to determine how much weight is actually needed. All tensioning devices need to be adjusted based on the size and length of the yarns.

- Before putting the yarns on the machine, coat them with beeswax. Also, before removing the rope from the machine, rub it down gently with a damp cloth. This helps the twist to set.

- When you see measurements for the sizes of ropes in books and drawings, remember that their thickness is measured by circumference (the distance around), not diameter (the distance across).

BLOCK MAKING

I recommend that you make your own blocks, except, perhaps, very small ones. Commercial blocks are expensive, and you'll need a great many. It's time-consuming, but not hard. Boxwood is the preferred material, though pear and holly work well, too.

Step One. Cut long strips to the width and depth of the desired blocks.

Step Two. Cut notches in the wood at regular intervals on the faces where the sheave holes will appear (**A** in Figure 9-13). These intervals should be slightly greater than the desired length of the blocks to allow for loss of material when cutting them apart.

Step Three. Mark the positions of the sheave holes and drill them using a small drill press (**B** in the figure). You can do this with a pin vise, if you prefer, but it's slow going.

Step Four. Cut the strop grooves and the grooves that represent the sheave openings with a small V-gouge. These grooves must be large enough to accommodate the size of line to be used with the blocks.

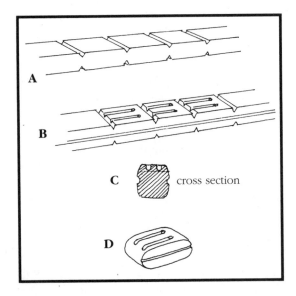

FIGURE 9-13. Block making.

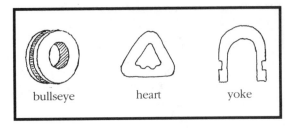

FIGURE 9-14. Miscellaneous wooden rigging items.

Step Five. Complete the shaping of the cross-section of the strip (**C**).

Step Six. Now cut off the individual blocks and sand them to their final shape (**D**).

Other wooden rigging items you might have to make include bullseyes, hearts, and yokes (Figure 9-14). You won't need many of these, and they're easy to make. Bullseyes and hearts are often used to set up stays and bowsprit shrouds. Yokes were used to set up forestays on the bowsprit. Don't bother to make deadeyes. It's easier to buy them, and the ready-made ones are probably better than you could do yourself.

HARDWARE

The most common hardware items modelers make are eyebolts, rings, bands for spars, and mast hoops.

EYEBOLTS. Commercial eyebolts are available only in certain sizes, they're expensive, and you'll need a lot of them. Add to that the fact that they're easy to make, and the choice is clear.

The easiest way to make eyebolts is to seize the end of a long piece of wire with round-nose pliers, bend it around one jaw,

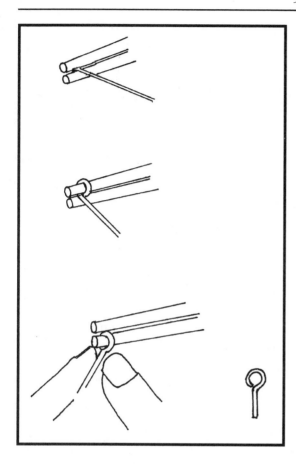

FIGURE 9-15. Making eyebolts.

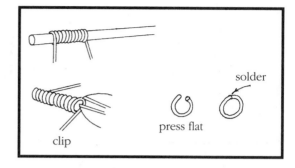

FIGURE 9-16. Making rings.

and then, with your thumbnail, bend the shaft perpendicular to the eye (Figure 9-15). Finally, cut off the shaft to the desired length. The fussy part is that you have to modify the points of the pliers in various ways depending on how large or small the eye needs to be. Two possibilities are to grind down the points or weld pairs of stiff wires to the jaws of an old pair of pliers. Experiment. See what works for you.

RINGS. Small rings are both essential and easy to make. Choose a piece of wire or rod whose outside diameter is equal to the de-

sired inside diameter of the rings. Wrap a piece of soft brass or copper wire around the rod as many turns as you need rings. Keep the turns tight against each other. Remove the "spring" from the rod, and clip along its length with small wire cutters. As you cut, the rings will be created (Figure 9-16). Since they'll be slightly skewed, press the rings flat so that the ends meet. You might want to seal them with a bit of solder to prevent the lines from pulling through. You can use rings as hatch cover lift rings, as jib stay hanks, or with eyebolts to make ring bolts — the uses are almost endless.

BANDS. Eye bands are everywhere in the rigging of a ship. The basic form is a narrow band with one or more tabs projecting from it.

Cut a brass strip the width and thickness of the band you need (it must be quite a bit longer than the circumference of the spar to allow for the forming of the tabs and the overlapping of the ends). Bend the strip in half, pinching the bend tight with pliers (**A** in Figure 9-17). Seize the band with the pliers at the base of the tab, and bend the two parts of the band outward (**B** in the figure). Next, wrap it around the spar and

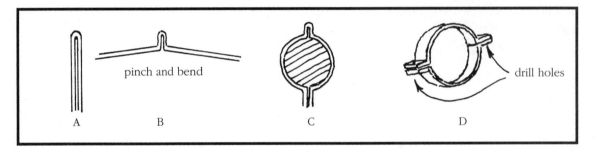

FIGURE 9-17. Making eye bands.

pinch the tabs closed at the other side (**C**). Cut off the excess. A bit of solder at each tab ensures that the tabs won't spread. Drill holes in the tabs large enough to accommodate a ring (**D**), and round the corners of the tabs with a file. Insert a ring in each tab, and solder the ring joint. Rings are necessary, since the edges of the tab holes can abrade the line.

Sometimes simple bands are required — for example, to strengthen the end of a boom where a gooseneck, rather than jaws, is fitted (Figure 9-18). Wrap a brass strip around the spar, and cut it to length so that there's a slight overlap. Then solder the band in place on the spar to ensure a proper

fit, and file down the little ledge formed by the overlap.

Other variations on simple bands include two bands joined to make a studding-sail boom iron, and a band and a piece of brass tubing soldered together to make a gooseneck socket (Figure 9-19).

MAST HOOPS. Mast hoops are used to secure the luff of a fore-and-aft sail. They're usually spaced about 18 inches apart. Late in the age of sail, these wooden hoops were sometimes iron rings. So, if the hoops on your model were iron rings, use the procedure above for making rings; if the mast hoops were made of bent wood, use this procedure.

Find thin wood that bends easily. Pine shavings are good. You can also use wood veneer tapes available at good hardware

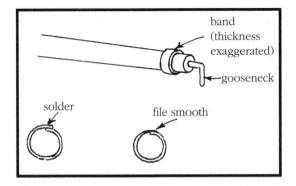

FIGURE 9-18. Booms were sometimes made with a gooseneck that fit into an eye on a mast band.

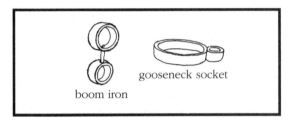

FIGURE 9-19. Band variations.

stores. Most of these tapes have self-adhesive backing, which would have to be removed before using for hoops.

Find a dowel with an outside diameter the same as the inside diameter of the hoops. (Hoops were all of the same diameter on a mast, regardless of any mast taper.) Soak a length of wood in household ammonia for a minute or so, and then wrap it around the dowel, overlapping it slightly (Figure 9-20). Secure the wood in place with string or a rubber band and let it dry thoroughly. Remove the resulting tube from the dowel and taper the ends just a bit at the overlap. Replace the tube on the dowel, glue it at the overlap, and secure it as before. (It's a good idea to wrap the dowel with waxed paper before gluing so the hoops don't stick to the dowel.) When the glue is dry, sand the tube smooth while it's still on the dowel. Then, with a sharp knife or razor blade, cut the hoops and slide them off the dowel. The support of the dowel is essential in cutting off the hoops without splitting the wood.

RIGGING

The subject of rigging is extremely complex, mostly because of the countless details that varied according to time and place. Read as much as you can on rigging your ship and other ships of the time. It's easy to become overwhelmed by the apparent complexity, but don't. Take it one step at a time.

Keep in mind that the principles are fairly simple. If you've built a rigged model from a kit, you've already served a useful apprenticeship. As your first scratchbuilt project, choose a ship with simple rigging and a comprehensive rigging plan. (It's best not to start with a full-rigged ship.) Conform to actual practice as nearly as possible. Shortcuts are rarely satisfactory, and sometimes lead to complications. And finally, take the time to learn and understand the sequence in which rigging should be set up, and know how to do a few basic ties.

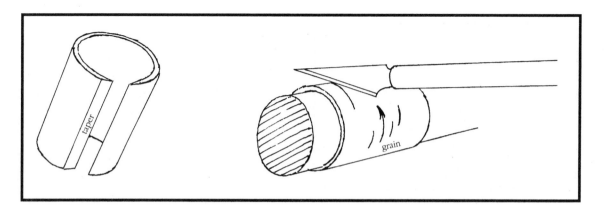

FIGURE 9-20. Making wooden mast hoops is easy.

SEQUENCE OF RIGGING

The sequence of rigging is critical. Be sure that all of your line is beeswaxed before using it. Start with the bare masts and bowsprit, and get the standing rigging on. I always do the bowsprit and jib boom first, starting with the bobstays and martingale stays and guys, and then doing the bowsprit shrouds and jib boom guys.

The next step is the lower foremast shrouds. Shrouds go on in pairs. A line forming two shrouds is passed around the base of the masthead and the ends are brought down so that they can be set up on two adjacent upper deadeyes. The two shrouds are seized together near the masthead. The first pair of shrouds goes to starboard, the next to port, and so on. Then go on to the other lower masts, following the same procedure. When all the lower shrouds are in place, do the ratlines, being sure to attach them with clove hitches, *not* with glue (Figure 9-21).

During your research, you might come across a recommendation to make up the shrouds and ratlines on a jig off the model, and then to attach the whole assembly. Don't do it. It makes it almost impossible to attach the shrouds properly to the masthead. Be sure to align the upper deadeyes. One way to control this alignment is to get all the shrouds over the masthead, set up the foremost and aftermost shrouds, and then set up the others on a line between them. Use a gauge made from stiff wire bent into the shape of an elongated U with the length of the base of the U equal to the desired distance between the top hole of the top deadeye and the bottom hole of the bottom deadeye, to control the spacing.

When attaching the lower deadeyes before rigging the shrouds, remember that the chain plates must be in line with the shrouds. You'll probably want to install these deadeyes before the masts are permanently in place. Put the masts (only the lower masts are necessary) in place temporarily, and tie a line to the heads where the shrouds will attach. Use this temporary line to establish a straight line from the head to the lower deadeye locations, and on to the chain plates (Figure 9-22).

The next step is to do the fore, main, and mizzen stays, then the topmast shrouds and topmast stays, and then the topgallant shrouds and stays. Last of all come the backstays, but here's a hint: Although it doesn't conform to the normal sequence of rigging, it's a good idea to rig booms and gaffs in place before rigging the backstays for the mast. Otherwise, it can be very difficult to get at the areas of the mast to which the boom and gaff are rigged.

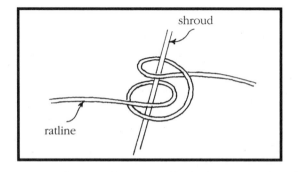

FIGURE 9-21. Clove hitches are used to bind the ratlines to the shrouds.

YARDS AND RUNNING RIGGING

Only now are you ready to deal with the yards and running rigging. Start with the

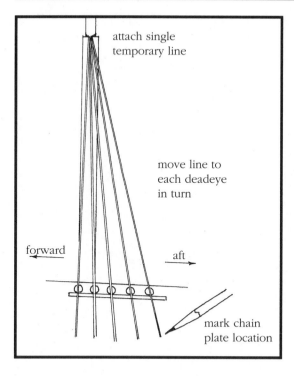

attach single
temporary line

move line to
each deadeye
in turn

forward

aft

mark chain
plate location

FIGURE 9-22. Aligning chain plates.

lower yards, working from fore to aft. Attach the yard to the mast, using parrals or hardware as appropriate, and then rig the halyards that raise and lower the yard. Finally, put on the topping lifts that support the ends of the yards and keep them level. When you're through with the lower yards, go on to the top yards, and then to the topgallant yards. Rig the braces, again starting low and working from fore to aft and up. Regardless of whether you put sails on your model, you might want to install the sail rigging (though it's quite all right to leave it off), and now's the time to do it.

Sooner or later, after running through one or more blocks, the ends of the running rigging come down to the deck and are tied in to belaying pins or cleats (Figures 9-23 and 9-24).

Be sure to do a neat job of the tying, and finish it off with a neat coil of line. On a model, the coil is separate from the rigging

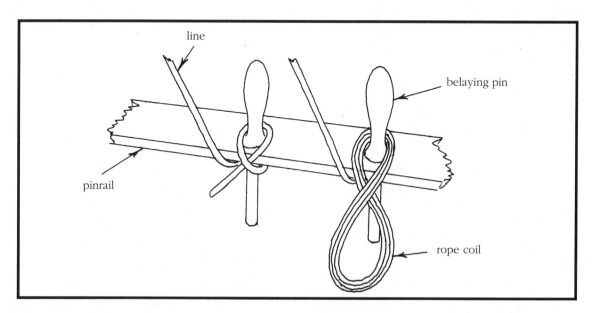

line

belaying pin

pinrail

rope coil

FIGURE 9-23. Belaying pin tie.

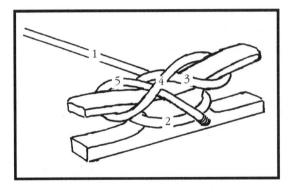

FIGURE 9-24. Cleat tie.

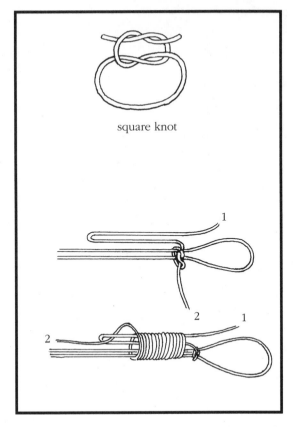

FIGURE 9-25. Square knots and seizing are everywhere in the rigging.

line. One way to do this is to take a length of well-waxed line and loop it a half-dozen or so turns around two headless brads driven into a board and spaced according to the desired size of the coil. Remove the coil and twist it into a figure eight or shape it into a long narrow loop. Secure the ends of the line with tiny drops of glue, and drape the finished coil over the belaying pin. If the line is coiled on deck, as it would be in the case of a cleat on the deck, the coil should be round.

KNOTTING AND SEIZING

Although it's beyond the scope of this book to demonstrate every method of knotting rope, you need to know how to tie a square knot and to do a proper job of seizing (Figure 9-25). You can do seizing any time you need to make a neat loop, such as in making the eyes in the shrouds and stays that go over the mastheads, in securing a block to the end of a line, or in tying a line to an eyebolt or a spar.

Though it looks complicated, seizing is easy: Tie a square knot around the two lines to be seized, leaving long ends. Loop end #1 back on itself as shown in Figure 9-25. Wrap the lines and the loop together with line #2 until the seizing is as long as needed. Then pull the end of line #2 through the loop of line #1, and pull on line #1 until the loop pulls line #2 into the seizing. Secure with tiny drops of glue, and cut off the ends.

DEADEYES

Finally, you need to know how to set up the deadeyes properly (Figure 9-26). Start by turning the end of the shroud around the

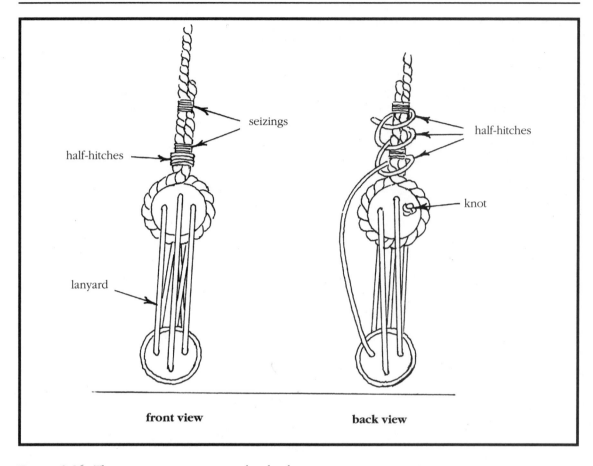

FIGURE 9-26. The correct way to set up the deadeyes.

deadeye and securing it with two or three seizings. The short end of the shroud should always be forward on the port side and aft on the starboard. Be sure that the short ends are all equal in length. Tie a knot in the end of the lanyard and pass it through the lower left eye of the upper deadeye so that the knot is at the back. Pass the lanyard through the front side of the upper left eye of the lower deadeye, through the back of the top eye of the upper deadeye, and so on. The last pass goes through the front of the lower right eye of the lower deadeye. Take the

lanyard up to the doubling of the shroud, and secure it with a few half-hitches.

A FEW FINAL WORDS

Always secure knots with glue, but use as little as possible. Don't use CA glue, since it makes the knots brittle. Gluing ensures that the knot won't slip, and makes it possible to trim the ends close to the knot.

Pay close attention to the proportioning of the rigging. Good plans specify the sizes of lines to be used, but you'll rarely have this advantage. Some reference books contain tables of rigging sizes for various purposes. The basic idea is that the size of rigging decreases as you go up. Lower shrouds and stays are proportionately heavier than topmast shrouds and stays. Running rigging is lighter than standing rigging, which means the main braces are much lighter than the main lower shrouds. If you can't find specific guidelines, consult plans of similar ships that provide such information, and use common sense. Just don't make the rigging too heavy overall. If you err, err on the side of lightness, but keep the proportions right. Excessively heavy rigging on a model makes it look crude and top heavy, even if the work is otherwise done well.

CHAPTER 10

Sailmaking and Flags

"No, no! We do not dip the sails in coffee and we do not starch them!"

— Clem Clew, master sailmaker, to a new apprentice who is a former model builder

Should you use sails on your model? That depends. Fore and aft sails almost always look good; square sails are another matter. Usually they hang lifelessly from the yards as if the ship were totally becalmed. There are ways to make them appear bellied out, but it never looks realistic. The exception is a waterline model mounted in a "sea," but it's not easy to carry this off. Furthermore, a full set of sails on a square rigger tends to hide much of the detail you've been so careful to create. Another consideration is that sails require much more rigging than if the yards were left bare.

The choice of whether to include sails is entirely up to you. Many modelers prefer to put the sails on fore-and-aft-rigged vessels, and to omit them from square riggers. A possible compromise for a ship such as a topsail schooner is to put on the fore and aft sails, and to show the square sails furled. Another interesting approach is to put on all sail rigging, and then string white line between the attachment points for the sails to represent their outlines. This is very effective, since it shows the approximate shape and location of the sails, yet permits an unhindered view of the ship's detail. Figure 10-1 shows typical sails with the components labeled.

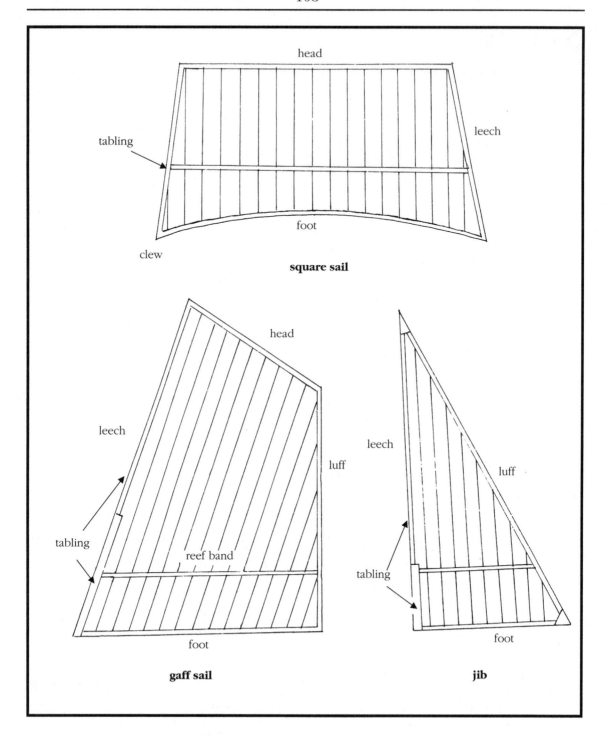

FIGURE 10-1. Typical sails.

Sailmaking

The first step is to find suitable material for the sails: as lightweight as possible, smooth, and tightly woven. Linen is best, but it's extremely difficult to find in a weight suitable for model sails. Egyptian cotton or batiste are also excellent. An alternative for small-scale models is to use model airplane silkspan, but I don't recommend it highly.

When making sails, be sure your hands are scrupulously clean. It'll save you the risky task of having to wash completed sails.

Step One. Trace the sail pattern onto your fabric so that the weave follows the line of the panels that make up the sail. (If you decide on furled sails on your model, don't make the sail patterns full size. No matter how fine your material, a full-size sail is impossible to furl so that it doesn't look bulky. Make the sail about one-third its normal hoist, but with full width at head and foot.)

Step Two. Now draw an allowance for the hem on the fabric (Figure 10-2). The hem represents what sailmakers call the *tabling*. Figure 10-3 shows how this was done in real practice. Since tabling is difficult to do on a small scale, use a simple hem (fabric turned over once) for these models.

If the scale is more than ¼"=1', you might want to try real tabling. If you do, mark the hem, allowing twice the amount of a simple hem since you'll be folding it over twice. The width of the tabling varied with time and place, but it was usually about 6 inches wide, except at the bottom of the leech tabling where it was wider (refer again to Figure 10-1). Keep the hem as close to scale as practical.

Step Three. Cut the sails out, being careful not to leave a ragged edge.

FIGURE 10-2. A typical sail pattern.

Step Four. If you're doing a simple hem, fold the hem over once and press it with an iron (on real sails, the tabling was always on the starboard side). If you're tabling, fold the hem over twice and press it.

Step Five. For a simple hem, set a sewing machine for a small zig-zag stitch and sew the hem. For tabling, use two lines of straight stitching instead of the zig-zag

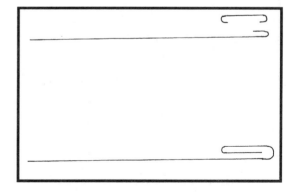

FIGURE 10-3. Methods of tabling.

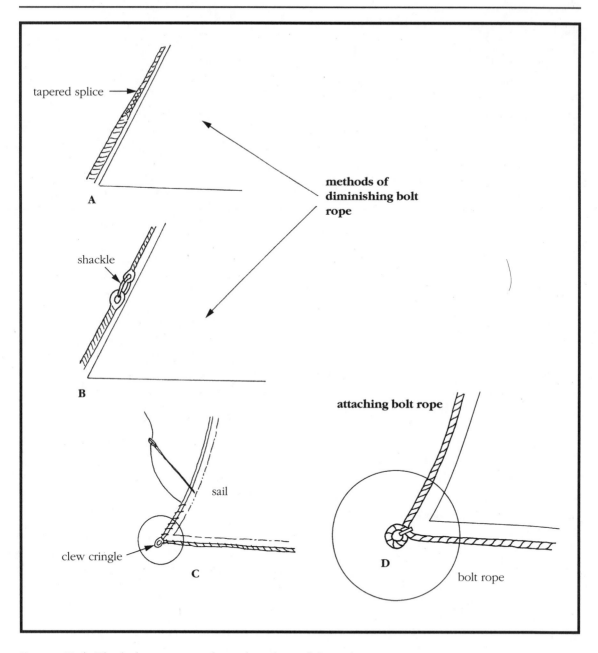

FIGURE 10-4. The bolt rope strengthens the edges of the sail.

stitch (some sewing machines have a feature that automatically doubles over the fabric and then sews the seam, in which case you don't need to fold the fabric manually and then press it down). Don't try to glue the hems down; it never looks good.

Step Six. Next you'll represent the reef bands and the seams joining the panels. If the model is to a large scale, you might want to sew lines of stitches to represent the panel seams, and to sew on actual reef bands. Needless to say, this requires extreme care and skill with the sewing machine. On smaller models (say, ⅜″=1′ scale or less), though, this tends to look bulky and out of scale. For most models, it's enough to draw lines representing the seams and bands on the sail with a 2H or H pencil. In showing the panels, remember that the standard width of sailcloth was about 24 inches. Keep the size of your panels to scale.

Step Seven. Attach the bolt rope (**C** in Figure 10-4). Use a waxed piece of brown line, appropriately scaled and long enough to go around the sail in one piece. (On real ships, the diameter of the bolt rope changed as it went around the sail: On the leech it was thicker near the foot of the sail; the transition was accomplished either by a tapering splice or by use of a shackle as shown in **A** and **B** of Figure 10-4. As you become more experienced, you might want to incorporate such niceties.)

Step Eight. Now sew the bolt rope to the port side of the edges of a fore-and-aft sail, or to the aft side of a square sail as shown in **C** of Figure 10-4.

Step Nine. Form the clew cringles at the corners of the sail. These are made by forming the bolt rope into loops. (The version shown in **D** of Figure 10-4 is somewhat

simplified.) This is a bit tricky, but stick with it. The clew cringles are used to attach the sails to spars.

Step Ten. Trim any ravelled threads from the hem.

Step Eleven. On small-scale models (⅛″=1′ and smaller), cut the reef lines to length and attach them to the reef bands with tiny drops of glue. On larger-scale models, it's best to replicate actual practice: Pull a line through the sail with a needle, knot it, and then cut it to length (**A** in Figure 10-5). A small drop of glue might be needed at the top of the line to make it lie flat on the sail (**B** in the figure).

Step Twelve. Press the completed sail with an iron.

Step Thirteen. Now you're ready to install the sail. Some people recommend dipping the sailcloth in tea or coffee to give it an "antique" look. Don't. Sails were not "antique," they were white (unless made with dyed cloth) and were bleached by the sun

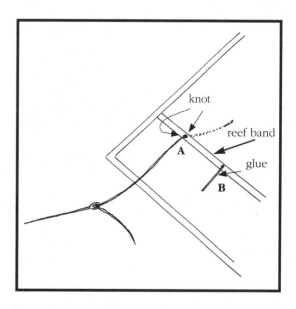

FIGURE 10-5. Attaching the reef lines.

and rain. The sails on an older model might have discolored with time, giving it a certain charm and testifying to the age of the model, but it's certainly not indicative of real sails.

If you decide, in spite of my advice, to show sails bellied out, you'll need to stretch the sails over carved forms to get the proper shape. Forms can be carved from blocks of pine or basswood. They should be checked for fit against the model and the stretched sail matched to the fitted form. Study drawings and photographs for the shapes taken by wind-filled sails. Before stretching the sail, soak it in starch or apply a thin coat of white acrylic paint. I recommend the paint — a starched sail is apt to be affected by moisture, thus spoiling its shape.

F‾LAGS

Flags are often considered a problem because of the presumed difficulty in forming them into a realistic shape. A real flag either hangs in folds or flaps in the breeze. It's not a flat, rectangular slab. There are a number of ways to deal with this, including making the flag from painted metal foil and then shaping it. The method I use relies on the fact that acrylic paints feel dry to the touch before they're completely cured, and even after curing remain pliable to a certain extent. This means you can paint the design on the cloth, let it dry to the touch, and then shape it into the desired folds. It'll hold the folds as it cures.

Step One. Begin by cutting a piece of thin, smooth fabric larger than the finished size of the flag.

Step Two. Tape the fabric to a drawing board over a piece of waxed paper at least as large as the fabric.

Step Three. Prime the cloth with white acrylic paint thinned to the consistency of light cream. When the fabric seems dry, turn it over and prime the other side. Priming prevents the colors from bleeding into each other when you paint the actual design. If the priming process raises fibers in the fabric, gently sand them away.

Step Four. Tape the dried cloth to the drawing board, and pencil in the design on both sides. Use a light box (or hold the fabric up to a window so the light shines through it) to ensure that the designs on both sides coincide exactly.

Step Five. You're now ready to paint the design, using thinned acrylic paint, good brushes, and a lot of patience. Be sure to leave a narrow white strip at the hoist to simulate the heavy binding used (Figure 10-6).

Step Six. When the flag is dry to the touch, cut it out of the larger piece of cloth. Because of the priming, no hem is necessary.

Step Seven. Now form the flag into the desired shape (Figure 10-7).

Step Eight. If the flag is large, you can insert tiny grommets into the hoist for the flag halyards. In most cases, though, it's enough to thread lines through the hoist

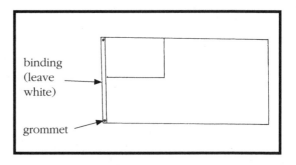

binding
(leave
white)

grommet

Figure 10-6. Flag binding.

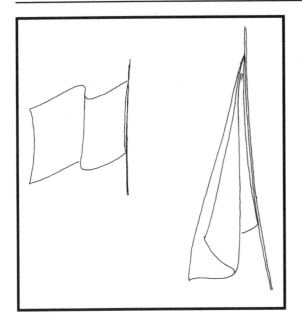

FIGURE 10-7. Draping the flag in a natural way.

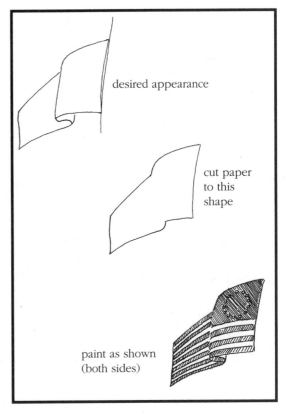

desired appearance

cut paper
to this
shape

paint as shown
(both sides)

FIGURE 10-8. A pre-shaped flag.

with a needle and then knot them to the halyard.

Another way to make flags on a small scale is to cut thin paper (cigarette paper is ideal) to the shape of the draped flag. Paint the design to fit the flow of the drape (Figure 10-8). This method is a bit tricky, especially in painting the flag convincingly, but it's very effective.

A final word on flags: Be certain that the design you're using is correct for the time and place. For example, an American flag must have the correct number of stars and stripes for the period. British flags in the eighteenth century were not quite the same as the ones we see today. And the French flew the tricolor during the revolution, the empire, and the republics, and flew the white and gold fleur-de-lys flag during the monarchy.

CHAPTER 11

Making Fittings and Furnishings

"Necessity is the mother of invention."

— Another anonymous savant

When you set out to make fittings and deck furnishings, there will inevitably be some items for which you can find no guidance, and you'll be on your own. Some of the items you might have to construct yourself are airports, anchors, cage masts, figures and carvings, lanterns, photo-etching patterns, searchlights, smokestacks, and ventilators.

Airports

Ready-made airports, often referred to as portholes, are widely available. You can,

however, easily make them for a small-scale model from pieces of brass tubing. The problem is in representing the glass. Trying to insert pieces of clear styrene is a hopeless task, and attaching the styrene to the outside of the tubing puts the "glass" too far back. The solution is to use clear casting resin.

Cut slices of brass tubing to the correct size and lay them out on a piece of waxed paper. Mix up a small batch of resin. Using a toothpick, drop the resin into each ring until it's not quite full. Let the resin cure. You'll be delighted by the realistic appearance of the completed airport (Figure 11-1). I've done this with tubing as small as $\frac{3}{32}$

---

inch in outside diameter. You can represent smaller airports by drilling holes of the proper diameter and painting the insides of the holes black. Don't try to paint a black dot on the surface. It won't look like an airport; it'll look like a black dot.

ANCHORS

Although you can get good commercial anchors at many hobby shops, you might have difficulty finding the exact size and pattern you need. Most suggestions for making your own anchors involve shaping or casting metal, which can be a hassle. You can use wood instead by making the shank and the arms in separate pieces (Figure 11-2).

Make the shank and arms from lengths of wood (**A** in Figure 11-2). Do not give the arms their final shape until they have been glued to the shank (**B** in the figure). I recommend CA or epoxy glue for this application. Drill the shank near the top for the anchor ring. After the arms are shaped, make the flukes from thin wood or plastic, and glue them in place (**C** in the figure). The stock is made from wood and drilled so that it can fit over the shank. You can make the bands around the stock from brass strips and solder them on or, on a small scale, from self-adhesive graphics tape (**D** in the figure). The shank, arms, and flukes should be painted black, and the stock stained. Glue the stock in place, and then fit the anchor ring. Be sure the anchor is correctly proportioned and is appropriate for the type of ship and period.

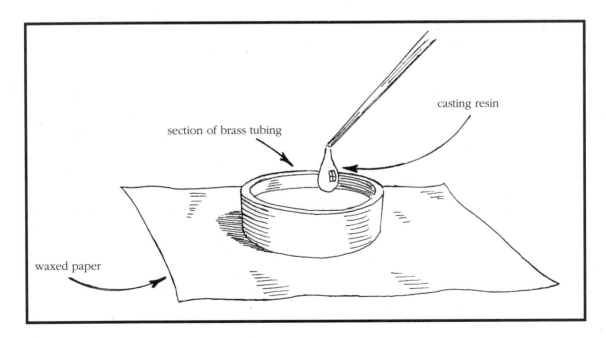

FIGURE 11-1. Making airports.

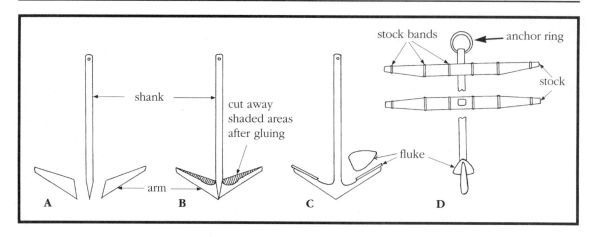

FIGURE 11-2. Making an anchor from wood.

CAGE MASTS

Quite a few modelers have refrained from building an early American battleship because they couldn't figure out how to construct the seemingly complicated cage masts (Figure 11-3). Too bad. It's a little involved, but certainly not difficult.

The construction is based on a series of risers that rotate 90 degrees as they ascend from bottom to top, half of them rotating clockwise, the other half counterclockwise. The problem is in controlling the many lengths of wire involved.

Step One. Turn a mold to the inside dimensions of the mast. Make sure the mold has flanges at top and bottom, and grooves where the mast bands will be located (Figure 11-4).

Step Two. Now notch the flanges. There will be one notch in each flange for each point at which a clockwise and counterclockwise riser intersect. Consult your drawing. The notches must be cut all the way to the body of the mold, and the

FIGURE 11-3. Cage masts are dramatic in appearance, and are simple to build. (Photo by Patricia Leaf)

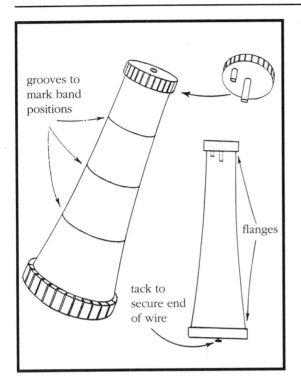

grooves to
mark band
positions

flanges

tack to
secure end
of wire

FIGURE 11-4. Cage masts are built up over turned molds.

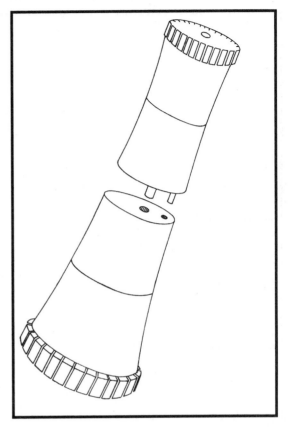

FIGURE 11-5. If the mast has flare at both top and bottom, you'll have to cut it at the least diameter.

notches on the top and bottom flanges must be aligned.

Step Three. Cut the mold and reattach it (necessary so that you can remove the mold from the finished mast):

- If the mast *doesn't* flare out toward the top, cut off the top flange and reattach it as shown in Figure 11-4.
- If the mast *does* flare, cut the mold where its diameter is the smallest and reattach the halves with dowels (Figure 11-5).

Step Four. Put a broad-headed tack in the center of the bottom flange to secure the wires you'll attach next.

Step Five. Now for the involved part. Follow along with Figure 11-6. Wrap one end of a long piece of soft brass wire a couple of times around the tack at the bottom flange. (It's best to work from a spool of wire so you don't have to splice, which would be messy.)

Step Six. Lead the wire through a notch in the bottom flange (**A** in Figure 11-6), make it turn 90 degrees to the appropriate notch in the top flange (**B** in the figure).

Step Seven. Pass the wire over the top

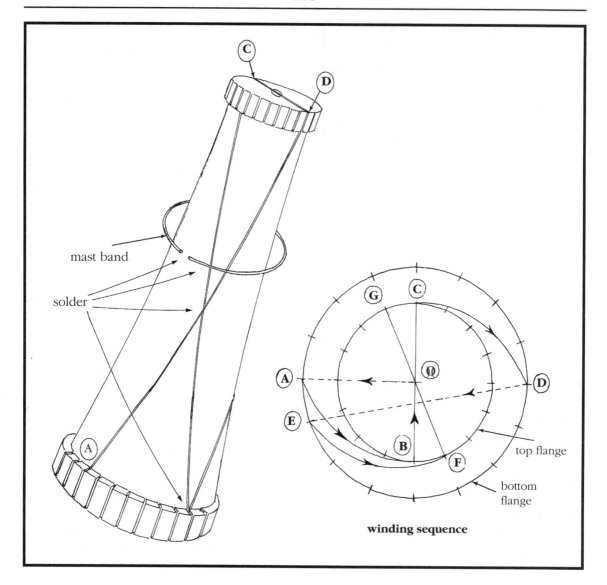

FIGURE 11-6. Wrapping the wires.

to the corresponding notch on the other side (**C**).

Step Eight. Run the wire to a notch 90 degrees away at the bottom (**D**), and then across and up again (**E**), and so on until all of the first set of wires are in place.

Step Nine. Now start again and do the same thing in the other direction.

Step Ten. When you're done, lightly solder each intersection. This sounds tedious, but it goes quickly and it's essential.

Step Eleven. Next, cut as many brass

strips as you need for mast bands and solder them in place as shown in the figure (use the grooves you cut in the mold as guides).

Step Twelve. Clean up the assembly by filing off any excess solder and removing any residual flux and dust.

Step Thirteen. Cut the wires at the top and bottom, remove the mold, and trim the ends of the wires.

Step Fourteen. Finally, add other details such as ladders, platforms, and the top according to your drawing.

FIGURES AND CARVINGS

A nice touch is to put a few crewmen on deck to define the scale of the model and to bring it to life. But carving such small figures, along with figureheads and other ornamental carvings, seems impossible to many modelers. The answer lies in the use of acrylic paste or an epoxy putty. Both materials enable you to mold a rough shape, carve it a little, add material if you cut away too much, carve again, and so on until finished. The advantage of the acrylic paste is that it gives you more working time for molding before drying.

To begin, make a stick figure (called an *armature*) from brass wire as shown in Figure 11-7 and solder the joints. You can now bend this stick figure into any position you like. When you've decided on a position, start building up thin layers of paste or putty. After the figure has dried, carve it to the shape desired. Yes, it takes skill and patience. The last step is to paint the figure. Figure 11-8 illustrates a seventeenth-century figurehead made using this process.

LANTERNS

The stern lanterns on older ships add significantly to the appearance and character of the ship (Figure 11-9). They look almost impossible to make, but they're really quite easy. The following method can be adapted to any shape of lantern. Simply analyze the design to determine where the separation point on the frame should be.

Step One. Make a wooden mold to the inside dimensions and shape of the lantern (**A** in Figure 11-10).

Step Two. On the mold, carefully draw the pattern of the lantern (**B** in the figure).

Step Three. Attach small brass plates at the top and bottom of the mold. They can be pinned temporarily until the vertical strips hold them in place. Cut brass strips of the proper width and thickness, and solder them to the brass plates at top and bottom, following the pattern you marked (**C** in the figure).

Step Four. Now put on the horizontal hoops, but make the center hoop in two parts, slightly separated as shown in the figure.

Step Five. When the pattern is complete, clean up the framework and then cut the vertical strips at the space between the center hoops. (A rotary power tool cutoff disc works well.)

Step Six. Pull the two halves off the mold.

Step Seven. Now cut segments of clear styrene plastic, fit them to the inside of the frame, and apply small dots of epoxy to hold them in place. (An alternative is to use casting resin to cast a solid filler in the shape of the original mold.)

bend brass wire components

solder

½

½

shape

build up layers

carve, file, sand

add costume

FIGURE 11-7. Making a figure by the armature method.

FIGURE 11-8. This complex 17th-century figure-head was sculpted using the armature method. (Photo by Patricia Leaf)

FIGURE 11-9. Well-built stern lanterns, like this one, add greatly to the appearance of a model. (Photo by Patricia Leaf)

Step Eight. Solder the two halves together with small dots of solder, or cement them with epoxy.

Step Nine. Finish the lantern with wooden base and finial (Figure 11-11).

PHOTO-ETCHING

The procedure for photo-etching is the same as that used in making tiny electronic components. It's invaluable in making delicate parts such as railings, radar screens,

and chain plates. All you do is make a precise large-scale drawing in black ink of the parts required, and then take the drawing to a company that does photo-etching. They will produce high-quality brass fittings using your drawing. The cost isn't high, considering the quality of the product.

SEARCHLIGHTS

Making searchlights is similar to making airports: Cut a piece of brass tubing for the

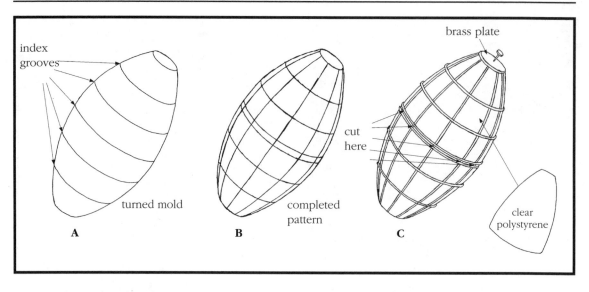

FIGURE 11-10. Making a lantern.

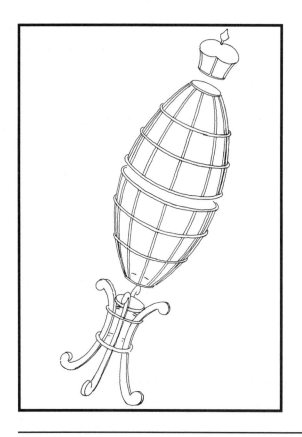

barrel of the light and, with CA glue or epoxy, cement a piece of wood to one end to form the back of the light (Figure 11-12). Don't shape this piece yet; leave it flat. Paint the inside of the piece of wood white. Now fill the tube with resin, and when it has cured, shape the back piece. The paint on the inside adds depth to the reflection in the "glass" and enhances the appearance of the completed searchlight.

I once used a variation on the resin technique to create the complex shapes of the lamps of a lightship. I filled a paper tube with resin, let it cure, and then turned the resulting rod on a lathe. The finished product was excellent (Figure 11-13), and it was even possible to illuminate it with a grain-of-wheat light bulb (available at any good hobby shop).

FIGURE 11-11. A finished lantern.

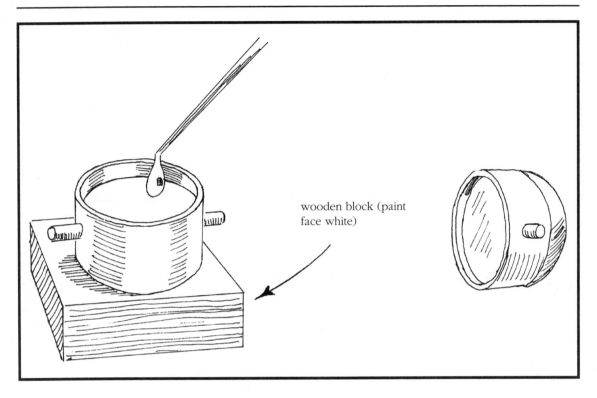

FIGURE 11-12. Making a searchlight.

wooden block (paint face white)

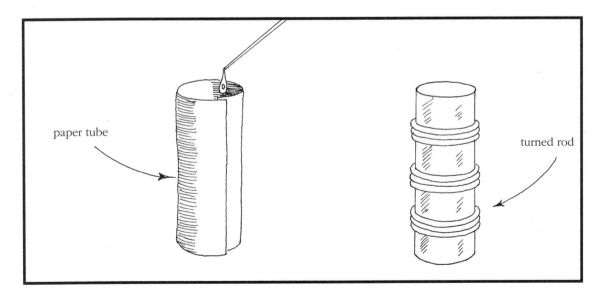

FIGURE 11-13. A lightship lamp turned from a casting-resin rod.

paper tube

turned rod

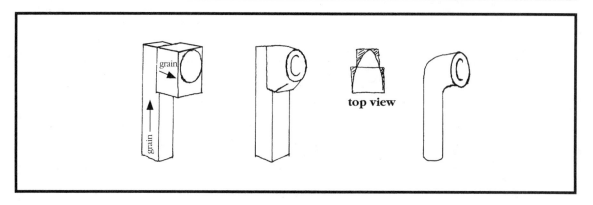

FIGURE 11-14. Carving a cowl ventilator from wood.

SMOKESTACKS

In making a smokestack, the worst things you could do would be to use a dowel or to carve it from a solid block of wood and then install it with the solid end of the piece showing. With the exception of some modern funnels with capped tops, smokestacks are open at the top, and their hollowness should be apparent. At the very least, hollow out the top of your block of wood to a depth that creates a reasonable illusion. Better yet, use a metal tube or, if the funnel isn't round or is too large for available tubing, form it around a wooden mold using a brass sheet or two or more laminations of thin card stock. Brass is clearly preferable, but if you're careful you can do a good job with laminated card stock. If you do this, don't pre-laminate the card. Laminate on the mold. The secret to a neat laminating job is to use a thin layer of glue; too much glue causes the card to buckle.

VENTILATORS

Figure 11-14 illustrates a good method for carving cowl ventilators. Notice the direction of the grain in the pieces. Gluing two pieces together in this way provides more control than attempting to carve the ventilator from one piece. After carving the outer shape of the ventilator, use a router to partially hollow it out.

CHAPTER 12

Displaying the Model

"Please! No! Don't stick your fingers in there!"

— Horrified ship modeler (who has failed to case
his masterpiece) to thoughtless guest

Sooner or later, your model will be finished and you'll want to display it proudly. Two elements are required: a stand of some sort to support the model, and a case to protect it from dust and curious hands.

STANDS

The type of stand you choose depends on the type of model and your personal taste. An ornamented cradle is fine for an eigh-teenth-century frigate, but totally out of character for a twentieth-century battleship. The classic mounting is a pair of turned brass pillars slotted for the keel. A screw passes through the bottom of the base board, through the pillar, and into the keel. This is elegant and simple, but might not provide adequate support for larger models. You might need additional side bracing (Figure 12-1).

Display cradles provide excellent support for the model and can be very simple or elaborately ornamented (Figure 12-2).

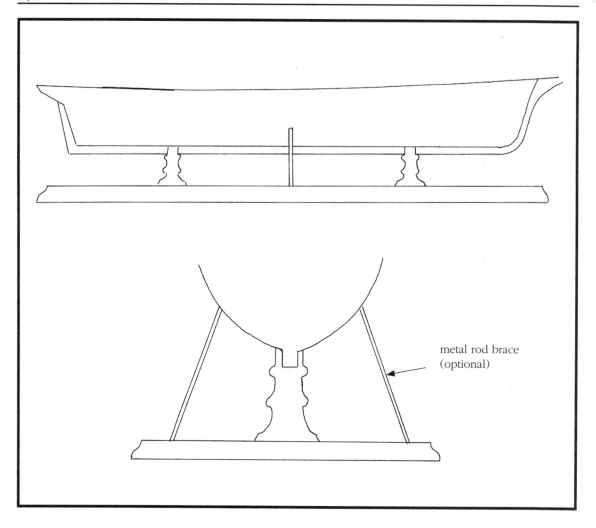

Figure 12-1. Mounting the model on pillars is traditional.

You can also adapt them to hold the model so that the waterline is level. Another effective stand, especially for modern models, is to represent the blocks used to shore up a ship in drydock (Figure 12-3).

You might want to build a model of a ship not quite completed, or just ready to launch. In such a case, why not mount it on a model marine railway? There are probably a million variations on these themes. Small models, for example, might be mounted on a single pillar. Just make sure the model is adequately supported and that the mounting is not anachronistic.

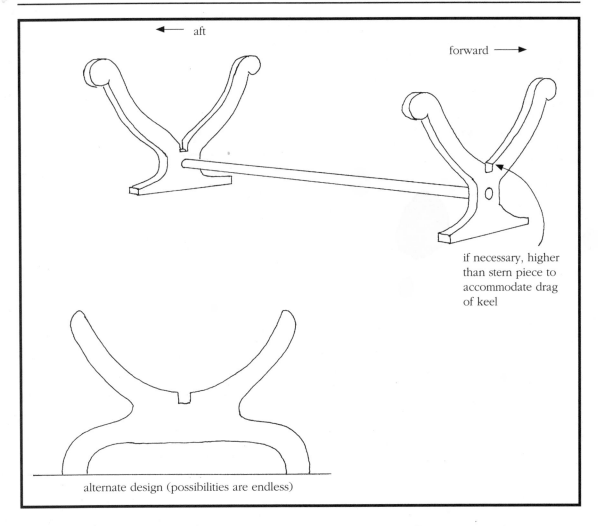

aft ←

forward →

if necessary, higher
than stern piece to
accommodate drag
of keel

alternate design (possibilities are endless)

Figure 12-2. Display cradles are practical and can be as simple or elaborate as you wish.

C̄ASES

Display cases can be simple glass or Plexi-glas boxes. They can be framed with wood-en moldings or with narrow brass angles. The bottoms of the cases can be plain var-nished wood, or covered with felt or velvet. If the bottom of the model is of interest, as in an unplanked hull, consider mounting a mirror under the model.

In any event, the case must be suited to the model. A case with heavy wooden moldings might detract from a small model.

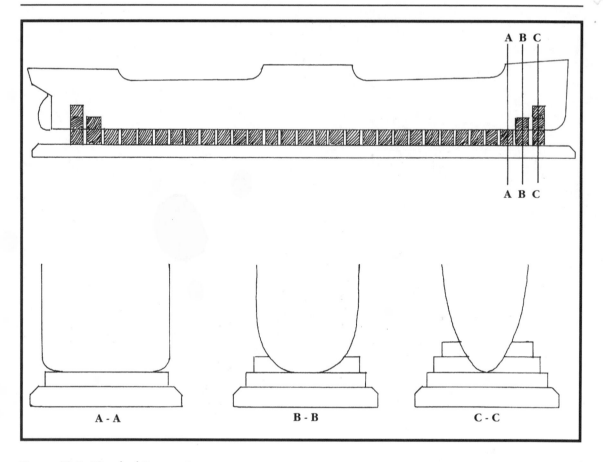

Figure 12-3. "Drydock" mounting.

You want the viewer to see the model, not the case. The case must be made so that the model can be easily installed and, if necessary, removed. Wood-framed cases have the advantage here, since they can have removable tops and side panels; large unframed glass cases are hard to handle. The glass should sit in grooves in the base to provide stability and to make a better dust seal. Wooden parts of the case should be oiled or varnished. Painted cases just don't look as good. Figure 12-4 illustrates some elements of case design.

The last bit of detail is an identification plate, perhaps a brass plate screwed to the outside of the base. Or, maybe you prefer to mount the plate on a slanted wooden block inside the case. Some modelers even do a brief history and description of the ship, frame it, and mount it inside the case.

Model display is an art in itself. A good display might not do much for a poorly made model, but it can certainly enhance a well-made one. Go to museums and model exhibitions and look for ideas.

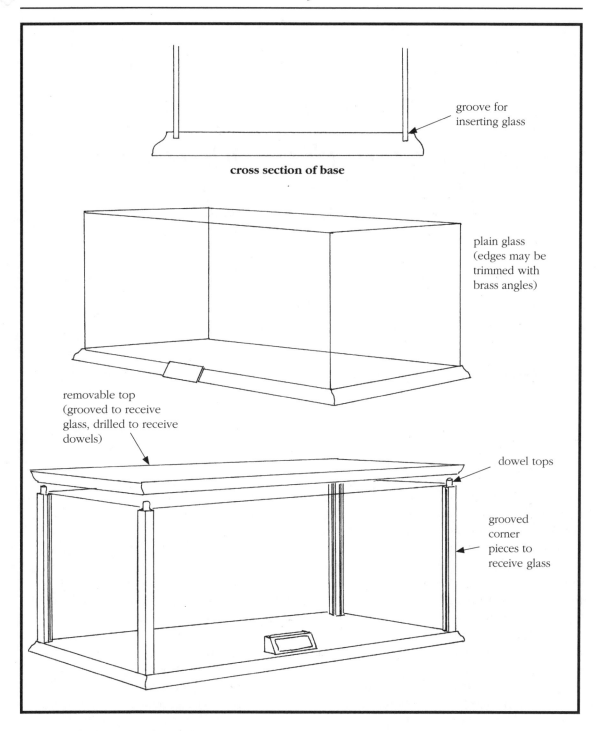

groove for
inserting glass

cross section of base

plain glass
(edges may be
trimmed with
brass angles)

removable top
(grooved to receive
glass, drilled to receive
dowels)

dowel tops

grooved
corner
pieces to
receive glass

Figure 12-4. Elements of case design.

CHAPTER 13

A Plank-on-Bulkhead Model from Start to Finish

"There ain't a tree big enough to make a solid frame. Bulkheads, you call 'em?"

— Sam Skeptic, suspicious shipwright

This chapter takes you through scratchbuilding an entire plank-on-bulkhead model. For your first planked model, it's best to start with a plank-on-*bulkhead*. True plank-on-*frame* (covered in the next chapter) is quite demanding: the skeleton is fragile and easy to deform if not properly supported.

The ship we'll be building is the *Brockley Combe*, a small British coasting freighter built in 1938. It's an attractive specimen of a small workaday ship, and well worth modeling. It'll give you a solid background in building a planked hull without the additional complications of framing and rigging.

A general layout drawing of the ship, with body plan, is shown in Figure 13-1.

The drawing is in many ways typical of the drawings available for modern ships. The lines are drawn to the outside of the planking. The ship was 171 feet long overall; if the model is built to ⅛"=1' scale, it will be about 21 inches long, a good size for a model.

Obviously, you won't be able to use the drawings in this chapter and the next to actually build the models. To do so, you'd have to enlarge the drawings and check the lines (and maybe enhance them). The drawings in Figure 13-1 are the result of my ef-

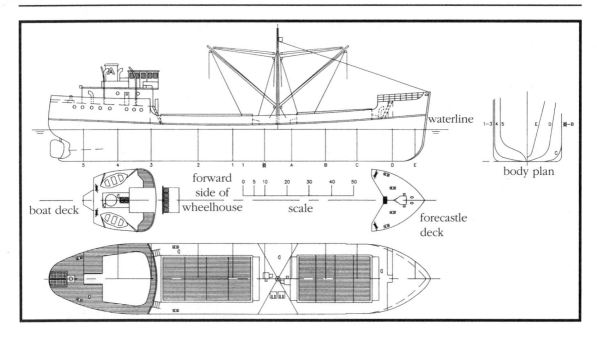

FIGURE 13-1. A working drawing for the *Brockley Combe* (for a full-page version of this drawing, see page 178).

forts in working from a small general arrangements drawing. The original drawing was cluttered with (to the modeler) irrelevant below-decks detail. There were no waterlines, body plan, or diagonals. I produced the body plan, largely through trial and error, with appropriately spaced sections. Still, there is much yet to be done if you were to attempt to use these drawings to build the model (see Chapter 2).

Preparing the Hull

I recommend using basswood for this model.

Step One. Begin by cutting out a backbone to the shape of the profile, extended to the level of the underside of the decks, and notched to receive the bulkheads (Figure 13-2). I used ³⁄₁₆-inch plank for both the backbone and the bulkheads.

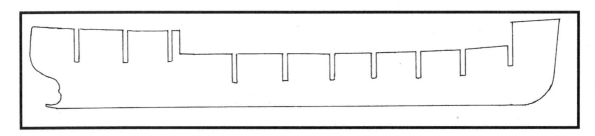

FIGURE 13-2. Backbone for the *Brockley Combe*.

A PLANK-ON-BULKHEAD MODEL FROM START TO FINISH

Step Two. Make patterns of the bulkheads from the body plan and cut them from plank stock, notching them to fit into the backbone (Figure 13-3). Notice the camber (¹⁄₁₆ inch at the midsection) on the tops of the bulkheads, and that temporary "timberheads" have been incorporated to support the bulwarks. Make these timberheads oversized, and cut them away *after* planking the bulwarks.

Step Three. Assemble the bulkheads to the backbone as described in Chapter 6.

Step Four. Set longitudinal stringers into the tops of the bulkheads for strength and rigidity. For a model of this size, one on each side will suffice.

Step Five. Now, bevel the bulkheads using a batten laid along them at various levels to determine how much needs to be taken off.

Step Six. Do the planking as described in Chapter 6, using carved blocks at the bow and stern. I used ¹⁄₁₆- × ¹⁄₄-inch strips for the planking, except at the turn of the bilge, where I reduced them to ¹⁄₈ inch. Notice that

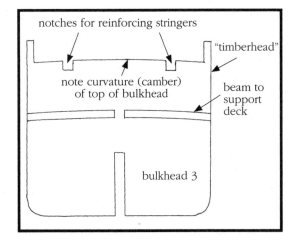

FIGURE 13-3. Typical bulkhead.

on the drawing (refer again to Figure 13-1), section E is shown even though the bulkhead itself is not used. Use this section to make a template as a guide in carving the bow blocks.

Step Seven. Before priming the hull, use wood filler or vinyl spackle to make the hull perfectly smooth. (This is a steel hull; you don't want the seams between the planks to show.) Don't forget the molding at the level of the main deck, or the narrow cutwater at the bow. If you wish, apply steel plating as described in Chapter 7.

Poop deck and forecastle deck

You should install the poop deck and forecastle deck athwartship bulkheads and lay the decks before planking the bulwarks. It's easier to fit them this way. Incidentally, if you don't want to go to the trouble of laid decks, you can use ¹⁄₃₂-inch commercial milled planking on this model. It's difficult to lay a deck with such narrow planks.

Step One. Install the main deck (which is steel, not planked). The deck will look something like an aligning jig, notched to fit around the timberheads. You'll have to add "deck beams" to the faces of bulkheads 3 and D. The ends of the main deck will rest on these.

Step Two. Put in the forecastle athwartship bulkhead. The forecastle has a straight bulkhead with a recess in the center for the ladderway. Bulkhead D is placed so that its aft side forms the back of the recess. The aft edges of the forecastle deck (also steel, not planked) will protrude beyond this bulkhead. The top edge of the hull planking should cover the outer edges of this deck. Right at the bow a little ledge will be formed by the intersection of the deck and the bow blocks. The small solid forecastle bulwarks at the bow will sit on this ledge.

Step Three. Now lay the poop deck, which is planked. Note that its forward edge is curved. Make a small piece with a matching curve and glue it to the main deck right under the poop deck's forward edge. The piece that forms the poop deck bulkhead will also form the rail across the forward edge of the poop deck, so make it high enough and keep the grain vertical. The vertical grain makes it easier to get a smooth bend.

Step Four. Glue the bulkhead to the poop deck and the piece you added below it, and then cut the openings in the rail for the ladders.

Step Five. Now you can install the main deck bulwarks. Their planks will extend into the forecastle and poop deck areas. It's best to cut the scuppers in the plank before putting them on.

Step Six. When the bulwarks are in place, cut away the temporary timberheads and glue thin triangular pieces to the insides of the bulwarks to represent the supporting plates found on the real ship. Like the small forecastle bulwarks already mentioned, the poop deck bulwarks at the stern will rest on a little ledge.

There's a complication here: a bit of inward slope at the stern, called *tumblehome*. This means you'll have to fit the bulwarks in this area using a pattern. Bend a piece of stiff paper around the stern so that it matches the tumblehome, and mark its intersection with the poop deck. This is the shape of the bottom of the bulwark. Cut the stern bulwark piece with the grain vertical.

Here's a tip: When you plank the bulwarks, make them a bit too high, and then cut them down to the correct height after they're securely in place.

Step Seven. When the hull is completely planked to your satisfaction, go ahead and paint it, and then stain the poop deck. Since you're not actually building the model now, I'll save all the information on the model's color scheme for the end of this chapter.

ADDING SOME DETAILS

Now we'll add some detail, starting with the propeller and rudder. Ready-made three-bladed propellers suitable for this model are widely available. That's good, because making your own propeller can be something of a chore. Now's also a good time to do the airports (covered in Chapter 11) and the hawse holes, located in the bulwarks at the bow. The openings shown in the forecastle deck lead to the chain locker.

DECKHOUSES AND HATCHES

The deckhouses and hatches come next. The large hatches are cambered and are covered with small planks about a foot wide. You can use $3/32$-inch commercial milled planking to simulate these planks.

First build the hatch coamings. These are simple wood frames lined to provide resting places for the hatch covers. They also have beams across them to support the covering planks. Each covering plank has a small lift ring at each end. Two smaller hatches are located on the starboard side of the mast between the main hatches.

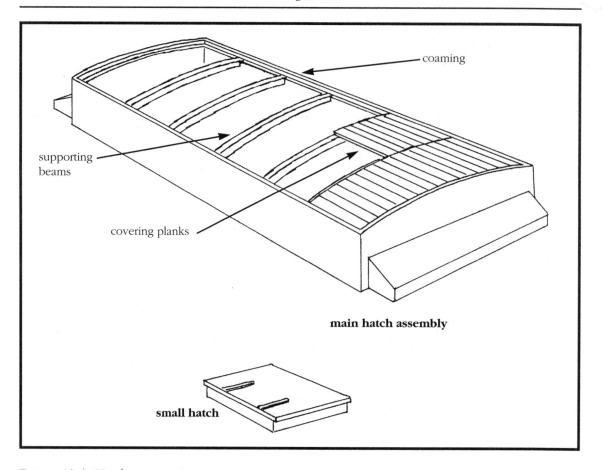

FIGURE 13-4. Hatch construction.

Before doing the main deckhouse, install the cap rails on the poop deck bulwarks and the rail across the front of the poop deck. The main deckhouse itself has curved bulwarks on the front and sides, and a number of airports. Cut the house from a solid block of wood (you'll have to carve out the bottom to fit the camber of the deck). Or, you can build it up with planks: Cut a base piece to the shape of the house, reduced by ⅟₁₆ inch all around. Cut a strip ⅟₁₆ inch thick and pin the base piece over it to produce the necessary camber (Figure 13-5).

Cut strips of ⅟₁₆-inch plank the height of the house with the grain vertical. Soak these in ammonia, secure them around the front and sides of the base piece, let them dry, and then glue them in place. The aft piece is straight.

Notice the small platform at the stern in Figure 13-1. This platform is actually a metal grating (which you can make from thin wire) within a wooden frame. Immediately forward of the platform is a capstan mounted on a block. You can turn or carve the capstan yourself (Figure 13-6).

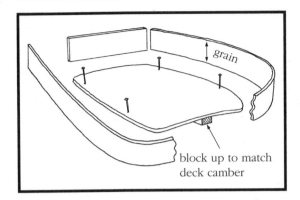

FIGURE 13-5. Constructing the deckhouse.

The top of the deckhouse forms the boat deck and extends beyond the house itself. This deck has a slight camber, but is flat in the middle where the wheelhouse and stack trunk rest. Part of this deck is planked, and part is plain steel.

On the top of the deckhouse are the wheelhouse and the trunk for the stack. The trunk is a simple box, which you can cut from a solid block. The wheelhouse is also a simple rectangular box with a flat top. Because of the windows, however, you'll have to build it up. Glue thin paper to the back of thin wood to keep the wood from splitting, and then carefully cut out the windows. On the real ship, the wheelhouse was made of vertically planked wood. Make it from the 1/32-inch milled wood you used for the poop deck.

Some overlaid framing surrounds the doors, which are located on each side of the house. Cement clear plastic to the inside of

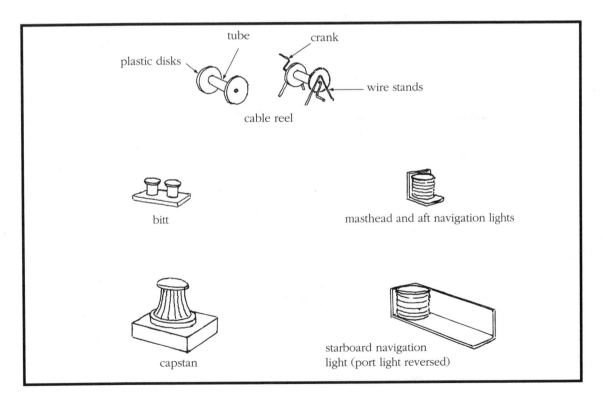

FIGURE 13-6. Miscellaneous fittings.

the house before you put the roof on. A single piece for each side is best. Now put on the solid rails that extend from each side of the wheelhouse part way around the edge of the boat deck. These rails have cap rails. It's best to cut these pieces with the grain vertical.

Behind the wheelhouse on the stack trunk are a small skylight, two cowl ventilators, and the stack. Cut a small block of wood to the shape of the skylight, stain the sides, and paint the slanted tops flat black. Cut the window pieces from index card stock, stain them, and then neatly cement them to thin clear plastic. Finally, glue the window pieces to the block. The illusion is excellent.

Now for the stack. On the top of the stack is a cap that you can make from two pieces of wood laminated cross-grain. It's best to cut the opening before shaping the cap (Figure 13-7).

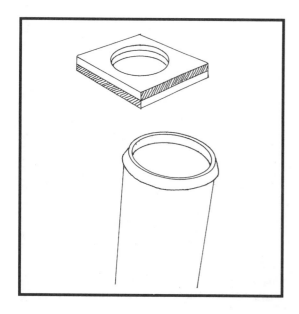

FIGURE 13-7. Making the stack cap.

WINCHES AND COWL VENTILATORS

The winches and the cowl ventilators come next. I recommend that you purchase cowl ventilators — you'll need six. As for the winches, there's a cargo winch at the base of the mast for each boom, and a mooring winch on the forecastle. The original drawing from which I developed Figure 13-1 was too small to discern any detail on these winches, so I've shown appropriate winches in Figure 13-8. You'll have to make the cargo winches yourself, but you can either make or purchase the mooring winch. Bluejacket Shipcrafters stocks a ⅛"=1' scale winch meant for a Liberty ship that's quite suitable. Figure 13-8 shows this winch and a simplified version of a mooring winch.

LADDERS AND RAILINGS

There are three ladders: two to the poop deck, and one to the forecastle deck. You can make the ladders from thin wood (Figure 13-9). Pin the side pieces to a board overlaid with waxed paper, and carefully cement the steps in place. Use a card template to keep the angle of the steps constant. Some people use a jig to assemble ladders, but I find it just gets in the way. (Commercial ladders of the correct size are also available, in wood or Britannia metal.) The ladders must have rails made of bent brass wire. Drill holes in the deck to accommodate the lower ends of these rails. Cement the upper ends of the poop deck ladder rails to the solid bulwarks with CA glue. Solder the forecastle ladder rails to the rails of the forecastle, and you're done with the ladders.

The forecastle is surrounded by open railing. Drill holes in the deck for the stan-

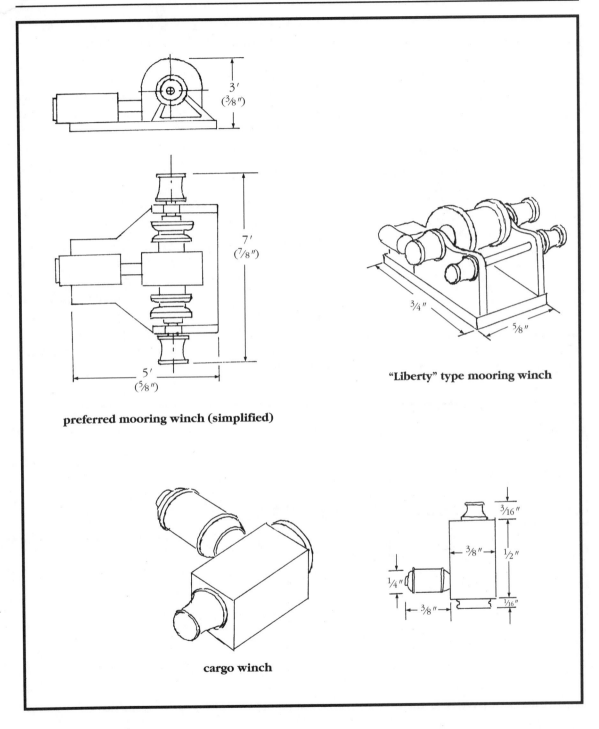

3'
($3/8$")

7'
($7/8$")

5'
($5/8$")

preferred mooring winch (simplified)

$3/4$"

$5/8$"

"Liberty" type mooring winch

$3/16$"

$3/8$"

$1/2$"

$1/4$"

$3/8$"

$1/16$"

cargo winch

FIGURE 13-8. Winches.

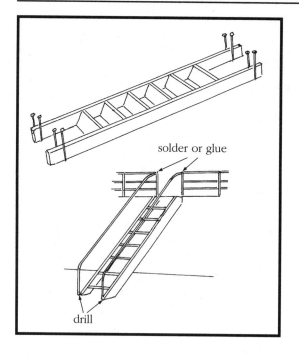

solder or glue

drill

FIGURE 13-9. Assembling a ladder.

chions (pieces of brass wire). Cut the wire a little longer than the required length, dip the ends in CA glue, and set the pieces in the holes. Cut a small block of wood to the desired height of the stanchions and use it to check that all stanchions are the same height. Use fine wire for the rails and solder it to the stanchions (Figure 13-10).

MAST AND BOOM ASSEMBLY

Now for the mast and boom assembly. Make this according to the instructions contained in Chapter 9. Or, you can make the mast of nested brass tubing, starting with a 3/16-inch outside diameter tube and working down.

The booms are attached to the mast with a gooseneck assembly (Figure 13-11). Just below each boom is a single block attached to the mast. There are two single

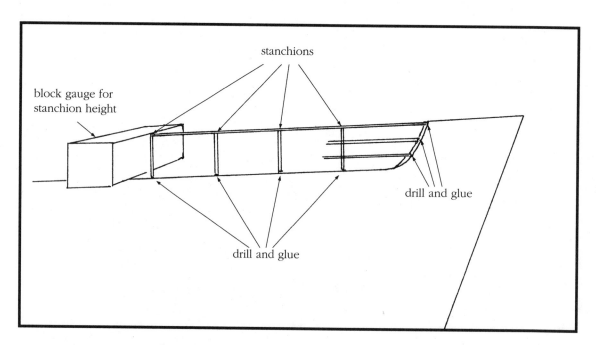

stanchions

block gauge for
stanchion height

drill and glue

drill and glue

FIGURE 13-10. Railing construction.

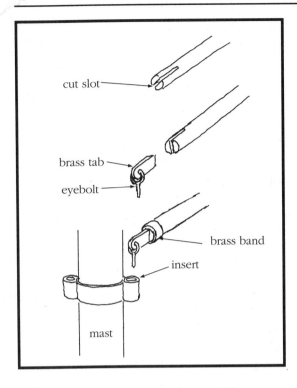

FIGURE 13-11. The gooseneck assembly.

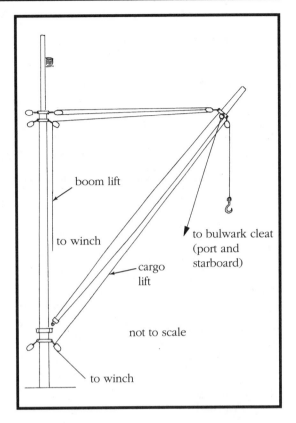

FIGURE 13-12. Mast and boom details.

blocks at the end of each boom: one for the cargo lift, and the other for the boom lift. There are also attachments (rings in eye bands) at the ends of the booms for the lines that control the booms' lateral motion. Two double blocks are located near the top of the mast for each boom lift. The booms and mast should be equipped with eye bands (covered in Chapter 9) to which the blocks are attached. The mast also has a band below the boom lift blocks for the mast guys, and a navigation light near the top. You can purchase Britannia metal fittings for the blocks and the light. Refer to Figure 13-12 for mast and boom details.

Mount the mast and boom assembly on the model, and rig the lifts. Each lift will have several turns around one winch drum.

Attach the guys to the mast band tabs (equipped with rings) with seized loops; secure the lower ends with turnbuckles attached to the top of the bulwarks with ringbolts. Figure 13-13 shows how to make a turnbuckle. Run lines from either side of each boom end to the mast guy attachment joints; these lines prevent the booms from flopping sideways.

DAVITS AND BOATS

Now let's do the davits and boats. The davits are made from pieces of brass wire bent to shape and tapered toward the ends.

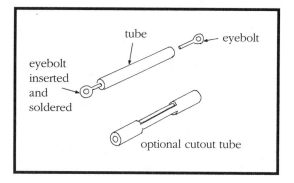

FIGURE 13-13. Making a turnbuckle.

This is hard to do with the size wire required, but try to ensure that the last ½ inch or so is tapered. Solder an eyebolt to the end of each davit, and solder a piece of wire shaped like a cleat to the sides. The davits are located at the edge of the boat deck. Drill holes in the poop deck for their lower ends, and secure them to the edges of the boat deck with narrow brass strips. A small double block is attached to the end of each davit, which is rigged to another double block equipped with a hook. Rig this tackle, but don't tie it down yet. Refer to Figure 13-14 for davit details.

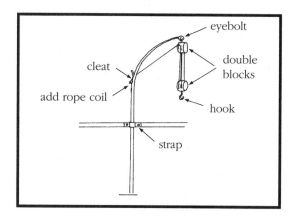

FIGURE 13-14. Details of the davits.

Carve the boats from blocks of wood. You can save yourself some work by making them as if they had their canvas covers on, as shown in Figure 13-15. Or, you can hollow them out and equip them with thwarts. Put an eyebolt near each end of the boats and mount them on little cradles. Hook the double blocks into the eyebolts, tighten up the tackle, and tie the ends to the cleats on the davits.

FINAL DETAILS

On the forecastle, main, and poop decks are a number of small bitts that you can purchase or make. You'll have to make the two cable reels on the forecastle and two on the boat deck. Make a drum from a piece of small brass tubing and two small plastic discs. Insert a piece of wire through the tubing and bend cranks into both ends. Bend wire into two inverted V's for the stand, and wrap line around the drum. Refer again to Figure 13-6 for illustrations of the bitts, reels, capstan, and navigation lights.

The last touches are the port and starboard navigation lights mounted on the bulwarks of the boat deck. Remember that the port light is red, the starboard light is green, and the masthead light is white. I suggest that you buy these lights.

COLOR SCHEME AND DISPLAY

As you build your model, paint each assembly individually before installing it. It's the only way to avoid getting paint where it shouldn't be, and it keeps you out of the trap of not being able to reach the items you want to paint.

Now for the authentic color scheme:

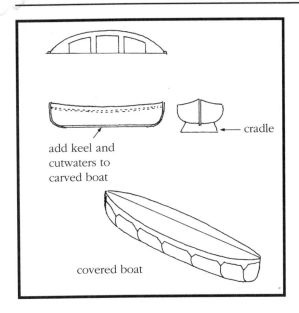

add keel and
cutwaters to
carved boat

← cradle

covered boat

FIGURE 13-15. Carve the ship's boats from blocks of wood.

Hull: Gray above the waterline with narrow white boot-topping; red below the boot-topping

Upperworks: Stone gray

Funnel: Yellow with a red letter *A*

Ventilators: Buff, with the insides of the cowls red

Hatch coamings: White

Hatch covering planks: Wood stain

Masts and booms: White

Winches: Gray

Ladders: White

Deck planking, wheel house, and hatch covers: Wood stain

This model provides a wonderful opportunity to try the weathering techniques described in Chapter 8.

This model could be appropriately displayed using a cradle or simulated drydock blocks. The case should be very plain — perhaps a plain glass box with brass-trimmed edges.

You'll notice I didn't describe every detail of building this little model. Some things are left for you to work out yourself — that's what scratchbuilding is all about. If you decide to try the *Brockley Combe*, good luck and have fun.

CHAPTER 14

A Plank-on-Frame Model from Start to Finish

"You don't really mean you're going to cover up that beautiful framework with planking!"

— Incredulous kibitzer

It's the ambition of almost every scratch-builder to build a plank-on-frame model of a rigged sailing ship. But don't get ahead of yourself. You're not ready for the *Flying Cloud* or the *Constellation*. Start small. A good first model of this type is the cutter-rigged *Lee*, an historic American ship that fought the British at the Battle of Valcour Bay on Lake Champlain during the Revolution. *Lee* has all the elements of a sailing warship, but with a hull that's easy to build and a simple, one-masted rig (Figure 14-1).

The lines on the drawing are to the inside of the planking. The vessel was 43 feet 9 inches between perpendiculars. At a scale

of 3/8"=1', the hull will be 16.4 inches long, not a bad size. I've provided interpretations of the original small-scale drawing, but there is yet more work to be done if you were to build the model yourself. If I gave you all the answers, it would almost be like building from a kit. This is scratchbuilding, remember. For details on preparing plans, see Chapter 2.

PREPARING THE HULL

We'll start with the keel/stem/sternpost assembly (described in Chapter 6). Use the deep-keel option shown in Figure 6-2. Don't forget to cut the rabbets (Figure 14-2). You'll want to make clamping and alignment jigs to keep the model steady and in the correct position. Clamp the keel assembly in the jig, and make the frames following the directions in Chapter 6.

Since the drawing (Figure 14-1) doesn't

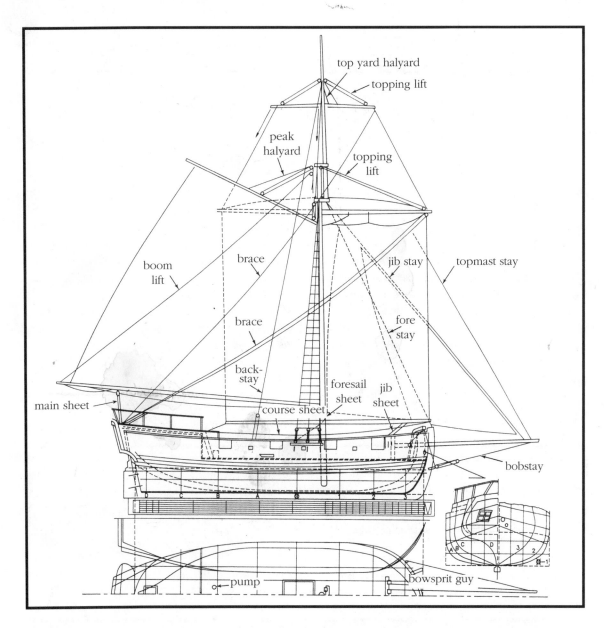

FIGURE 14-1. Working drawing for the *Lee* (for a full-page version of this drawing, see page 179). (Based on a drawing from *The History of the American Sailing Navy*, W. W. Norton & Co.)

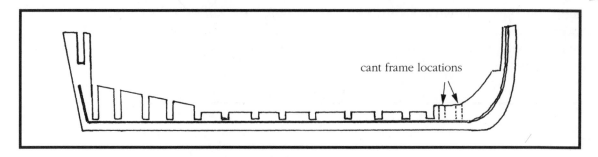

FIGURE 14-2. The keel assembly.

show many sections, develop a few more so that you have enough frames to support the planking. There should be one additional frame between each section shown on the drawing. In particular, there should be a frame at the location of the quarterdeck step, and one at the stern perpendicular. The latter frame is partly solid, since it serves as a base for the transom framing. You'll also need to develop some cant frames at the bow. The locations of these new frames are shown in Figure 14-3.

Before setting the frames in place using the aligning jig, glue a ¼-inch-thick piece to each side of the cutwater so that it fits ex-

actly against the inner rabbet. These pieces, known as *stem doublers*, provide a solid gluing surface for the bow ends of the planks; the rabbets alone aren't enough. At the stern, install the aftmost part-solid frame, and glue the central supports for the transom to each side of the top of the stern-post, butted against the frames. See Figure 14-4 for details of these pieces.

You're now ready to install the frames. When they're in place, put spacers between them below the levels of the decks. Remove the aligning jig. Now put in the beam shelves (remember that the top of the shelves is at the level of the undersides of

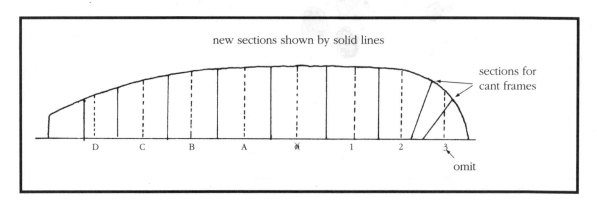

FIGURE 14-3. Additional sections have to be plotted so that you can make enough frames.

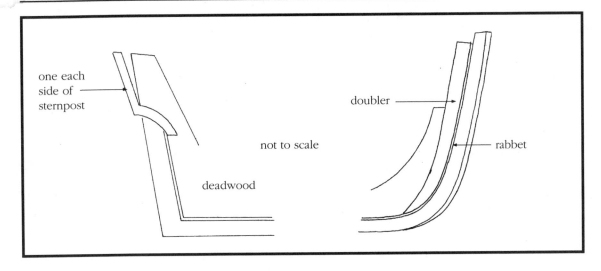

FIGURE 14-4. Transom supports and stem doublers.

the deck beams). Because of the extreme curvature at the bow, you'll need to carve pieces to fit to complete the shelves.

Now for the deck beams. They need to be thick enough to extend from the top of the beam shelves to the underside of the decks. They're cambered, of course. The curve on each beam is the same, regardless

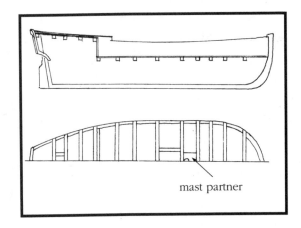

FIGURE 14-5. Deck beam layout.

of its length. Establish the required curve at the midsection and use this as a template for cutting all the beams. You'll need more beams than there are frames. A layout for the beams is shown in Figure 14-5.

Refer again to Figure 14-1. Be sure to include the framing for the hatch, skylight, and mast partners. There must be a deck beam just in front of the frame, which defines the quarterdeck bulkhead.

Framing the Stern

For many modelers, framing the stern is the most difficult part of framing the ship. Don't be discouraged. Take it slow and refer to Figure 14-6 throughout the following discussion.

You've already attached the first transom supports to the top of the sternpost. You'll need another pair of inner supports (**A** in Figure 14-6) sloped to match the

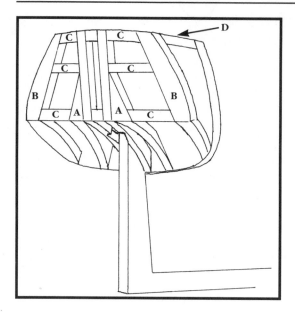

FIGURE 14-6. Transom and counter framing.

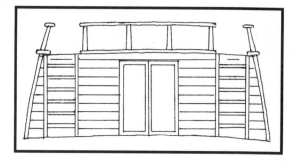

FIGURE 14-7. Quarterdeck bulkhead.

frames of the windows, and intersecting the first two supports. The outermost supports (**B** in the figure) actually serve the purpose of half-frames. They must be wide enough so that you can carve them to a shape that fairs in with the frames just forward of them. It's best to make them in two parts: one framing behind the transom, and the other framing the counter.

Notice in Figure 14-6 the spacers (**C**) between the inner and outer supports and between the transom framing and the frame just forward of it (**D**). The transom is slightly curved. If you've made the supports substantial enough, you can sand this curve into the transom supports after the framing is finished. The last step is to cut the opening for the rudder into the innermost supports. Framing the stern is a tricky operation — you have to make the pieces accurately, and the assembly strong.

Now's a good time to do the decks (covered in Chapter 6). Put in the facing for the quarterdeck bulkhead. Although it's not shown in the original drawing, this bulkhead probably had a door. This is the kind of educated guess you'll frequently have to make when working from old drawings. Frame up the bulkhead as shown in Figure 14-7.

PLANKING THE HULL

With the stern framed, it's time to plank the hull (covered in Chapter 6). Figure 14-1 shows two wales, which on the original vessel were strakes of greater thickness than the others. Put on these wales, with the intervening normal plank, first. Then put on the garboards and the sheer strakes, which on this model run at the level of the underside of the main deck cap rail. Plank the bulwarks solid. Before cutting the gunports and oarports, frame them on the insides of the bulwarks as shown in Figure 14-8. Now cut the gunports, oarports, hawses, and hole for the bowsprit.

The *Lee* would not have had port lids. However, the insides of her bulwarks

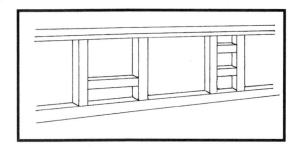

FIGURE 14-8. Gunport and oarport framing.

should be planked. Such planking was normal for warships.

With the hull planked, make and install the transom and plank the counter (the curved section below the transom). Refer to

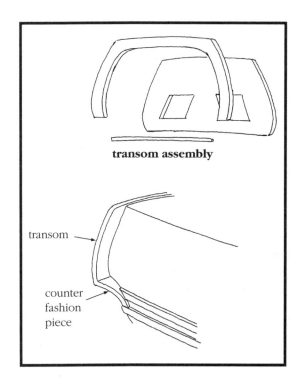

transom assembly

transom

counter
fashion
piece

FIGURE 14-9. Transom and counter.

Figure 14-9. First plank the counter to the edge of the hull planking. The transom is made up of two layers: a base piece that includes the openings for the lights, and a trim piece. The transom protrudes beyond the sides of the ship. You'll have to add fashion pieces to the sides of the counter to complete the shape.

Because of the slope and curvature of the transom, you'll have to develop its true shape (covered in Chapter 2). Then cut the transom pieces. The base piece should be $3/32$ inch thick, and the trim piece $1/32$ inch thick. Install the base piece first, and then put on the trim piece. If you try to do the whole thing at once, it'll be difficult to make it curve to fit. The lower edge of the transom overlaps the top edge of the counter planking. Add the narrow molding at the intersection of the transom and counter, and at the framing of the windows. Don't forget to glue clear plastic to the inside of the transom before installing it. The last step is to add the fashion pieces at the ends of the counter. Carve them from a block of wood and then fit them to the hull.

The cap rails come next. These go on top of the bulwarks, and are a foot wide. Since you can't bend them, they will have to be made in several pieces so that the grain of the wood of each piece follows the curve as nearly as possible. Next, put on the molding that extends the line of the maindeck rail to the stern.

Now make and install the rudder and tiller assembly (the gudgeons and pintles are covered in Chapter 7). Cut holes in the counter and in the quarterdeck to accommodate the rudder head (Figure 14-10 shows the rudder head and tiller). The shape of these holes is trapezoidal, with the wider base curved and outboard.

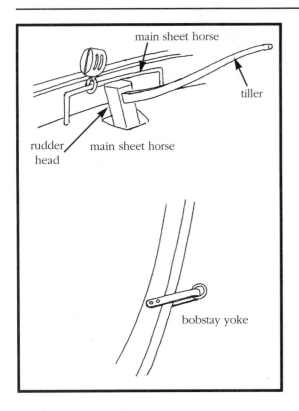

FIGURE 14-10. The horse is for the main sheet, and the yoke supports the bobstay at the cutwater.

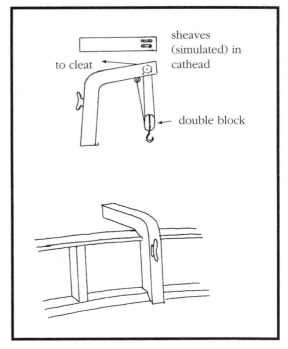

FIGURE 14-11. Cathead details.

DETAILS AND DECK FURNISHINGS

Now it's time to paint the hull (the color scheme is included at the end of this chapter). When this is done, make the channels for the shrouds, the little steps a bit abaft of the channels, the quarterdeck rails, and the catheads (Figure 14-11).

A railing runs along the forward edge of the quarterdeck, extending between the inner edges of the ladders. Remember that *Lee* was built in primitive conditions. The stanchions for the quarterdeck railings were probably square in cross-section, tapering slightly toward the tops (no fancy turnings). Set them into holes in the cap rail.

There are very few deck furnishings: a main deck hatch, a skylight on the quarterdeck, bitts to support the heel of the retractable bowsprit, two ladders, and two pumps. To make the ladders, follow the directions provided in Chapter 13, but omit the rails. Figure 14-12 illustrates some of the deck furnishings.

Although not shown on the original drawing, there would have been a fife rail at the foot of the mast, and pinrails along the insides of the bulwarks (Figure 14-13). These rails held the belaying pins to which the running rigging is secured.

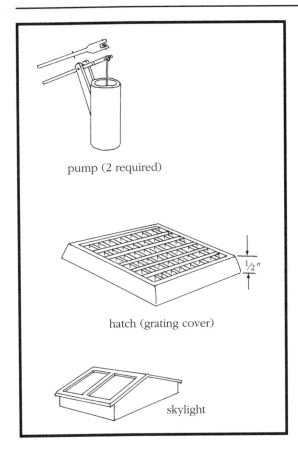

pump (2 required)

hatch (grating cover)

½"

skylight

FIGURE 14-12. Deck furnishings.

You'll also need a horse (a low rail for the attachment of the main sheet), a metal yoke on the cutwater for the inboard end of the bobstay, cleats to secure certain lines, and eyebolts or ringbolts in the deck for the ends of the running backstays and the jack-stay. Figure 14-10 shows the horse and yoke.

You can also make the anchors at this time, according to the instructions provided in Chapter 11. Figure 14-14 shows the correct pattern for the anchors. Figure 14-11 shows how to rig the catheads. The anchors hang from the hooks on the double blocks.

Making and installing the bowsprit is simple (Figure 14-15). It has a square heel to fit in the supporting bits, and then transitions to round. The bobstay, bowsprit guys, and topmast stay are secured to a shoulder near the end of the bowsprit. The other end of the bobstay is secured to a yoke on the cutwater by a block and tackle, the end of which terminates at a belaying pin at the bow. The bowsprit guys run to a block and tackle secured to ringbolts just below and abaft of the catheads. The ends of these tackles also terminate on belaying pins.

Guns come next. *Lee* was pierced for eight guns, but only carried six carriage guns. According to contemporary British sources, these were four 4-pounders, one 9-pounder, and one 12-pounder. I checked the measurements, and there's no way that a 12-pounder could have been mounted on this ship. (So much for sources.) Let's make four 4-pounders and two 9-pounders. Figure 14-16 shows the measurements for both, and also shows how the guns were rigged. The drawing shows standard proportions for guns and carriages. It might be necessary to raise the height of the 4-pounder carriages so that the guns will center in the ports. The 9-pounders will go in the foremost ports; the aftmost ports will not be filled.

RIGGING

At last we're ready to attack the rigging. *Lee* was cutter rigged. Chapter 9 covers the basics of rigging, but there are some peculiarities inherent in the cutter rig, which uses a running square sail. The December 1984 issue of the *Nautical Research Journal* (Volume 30, Number 4) contains a comprehen-

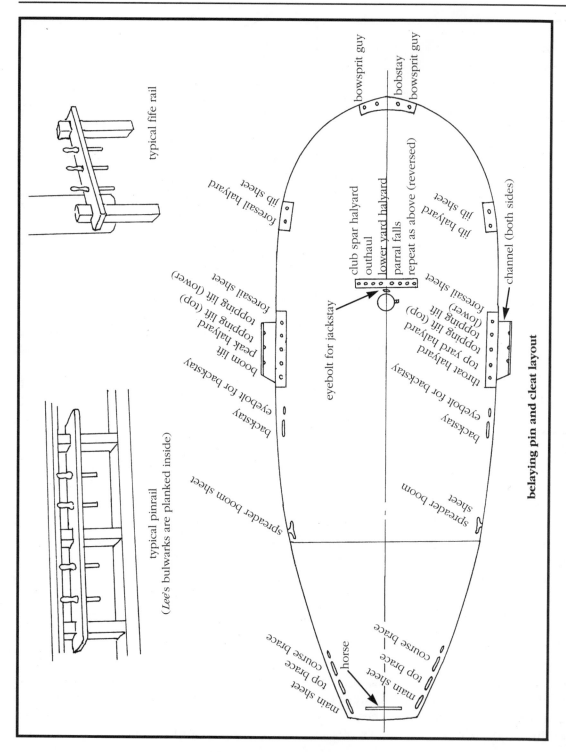

typical fife rail

typical pinrail
(*Lee's* bulwarks are planked inside)

bowsprit guy

bobstay

bowsprit guy

foresail halyard

jib sheet

jib halyard

jib sheet

channel (both sides)

club spar halyard
outhaul
lower yard halyard
parral falls
repeat as above (reversed)

foresail sheet

eyebolt for jackstay

boom lift
peak halyard
topping lift (top)
topping lift (lower)
foresail sheet

throat halyard
top yard halyard
topping lift (top)
topping lift (lower)

backstay
eyebolt for backstay

backstay
eyebolt for backstay

spreader boom sheet

spreader boom
sheet

main sheet
top brace
course brace

horse

main sheet
top brace
course brace

belaying pin and cleat layout

FIGURE 14-13. Pinrail and fife rail.

SHIP MODELING FROM SCRATCH

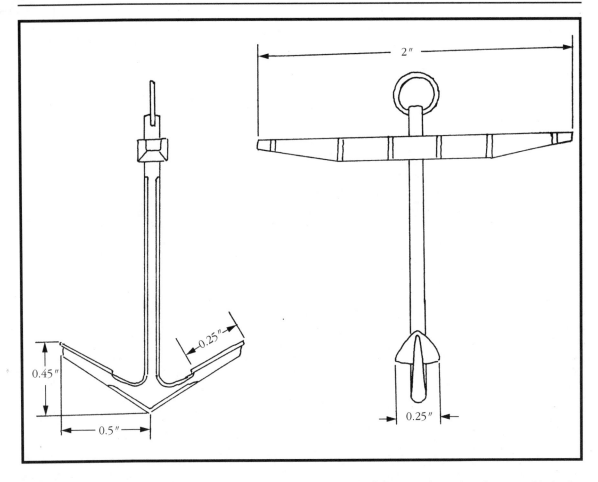

FIGURE 14-14. This is a typical British anchor from the period of the Revolution, and is most likely the type used on the *Lee*.

sive, well-illustrated article about this type of rig.

The mast structure is conventional — equipped with a boom rest, hounds, cross-trees, and trestletrees. Figure 14-17 shows the details of the main top, including the locations of the tackle. Figure 14-18 shows how the fore and jib stays and shrouds are set up. Figure 14-19 shows how the gaff halyards and the boom lift are rigged to the top. Figure 14-20 shows the rigging of the

main sheet, which controls the movement of the boom.

The top mast has a shoulder at which the topmast shrouds, topmast stay, running backstays, and blocks for the lifts are attached. A sheave is included just below this shoulder for the top yard halyard. The boom, gaff, and yards are also standard. The lower yard and the aft end of the boom are equipped with footropes.

A feature unique to the running square

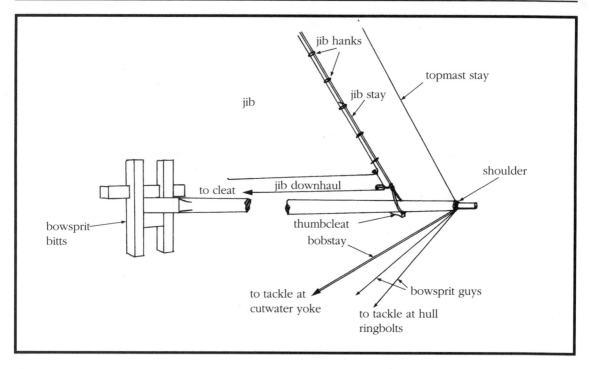

FIGURE 14-15. Bowsprit details.

sail is the vertical jackstay and the spreader boom at the bottom of the main course. The jackstay is a heavy rope that's secured by a loop over the trestletrees, and then goes vertically down the front of the mast and is set up with deadeyes to an eyebolt in the deck. The lower yard is attached to the jackstay with a rope traveller. On each side of this traveller is a smaller one through which a rope parral passes. The ends of the parral go down to pins in the fife rail. This made it possible to slack off the yard from the mast, and to brace it around radically.

Outboard of the parral travellers are the lower yard halyards. The spreader boom was not secured to the vertical jackstay, but was lashed at varying angles to eyes in the cap rails. The lower and top yards are fitted with braces, and the spreader boom has sheets. Also, the topmast shrouds are not secured with deadeyes, but are set up on single bullseyes.

I suggest that you spread the gaff and head sails, and either show the square sails furled or omit them entirely. If you do this, lower the top yard to the level of the mast cap. The spreader boom rests on the tops of the bulwarks with the sail furled to it. The rigging of the main course merits special attention. The middle third of the head of the course is lashed to a club spar, which is hoisted by halyards to the foreyard. The head clews are spread to the ends of the yard by outhauls. The clews at the foot are

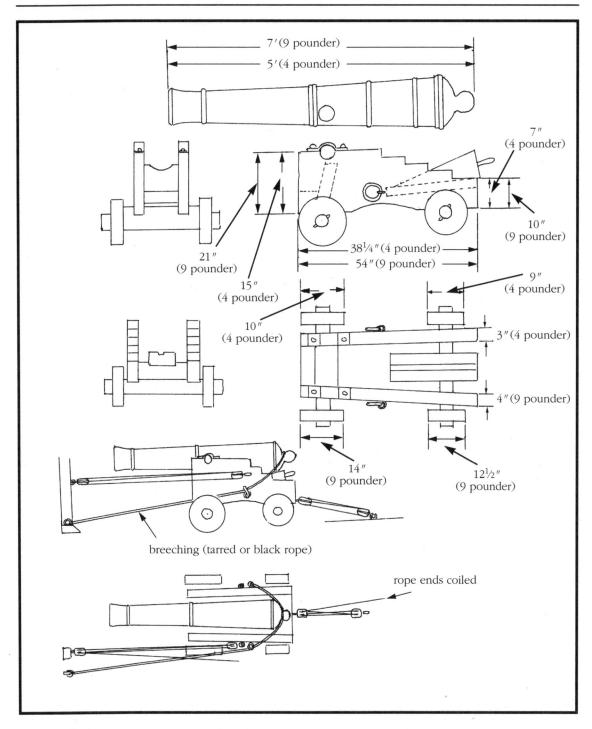

FIGURE 14-16. Gun construction and rigging.

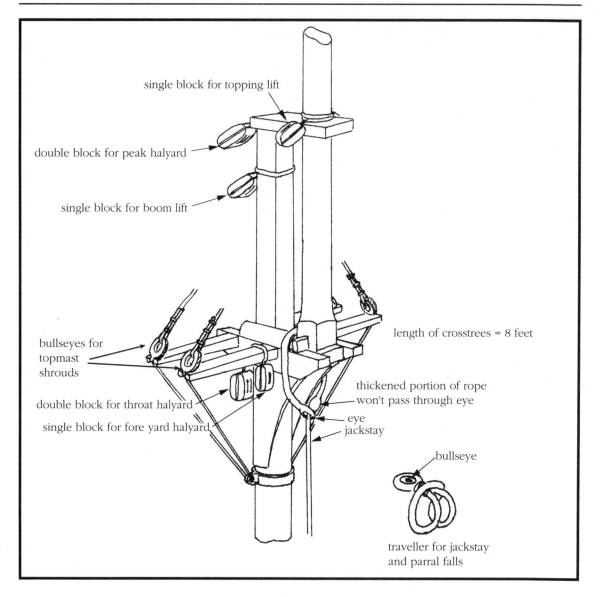

single block for topping lift

double block for peak halyard

single block for boom lift

length of crosstrees = 8 feet

bullseyes for topmast shrouds

double block for throat halyard

single block for fore yard halyard

thickened portion of rope won't pass through eye

eye

jackstay

bullseye

traveller for jackstay and parral falls

FIGURE 14-17. Main top details.

spread by sheets that pass through blocks at the ends of the spreader boom. Figure 14-21 shows how all this works.

The gaff sail is loose-footed — that is, only the cringles at the foot are attached to the boom. The head of the sail is lashed to the gaff. Refer to Chapter 10 for sailmaking instructions. The headsails are attached to the stays with small metal rings called hanks. The lower end of the forestaysail luff

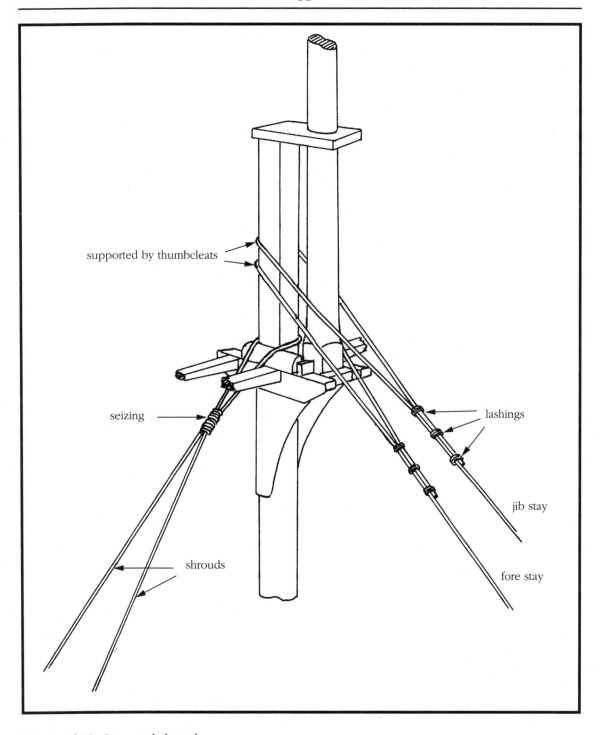

supported by thumbcleats

seizing

lashings

jib stay

fore stay

shrouds

FIGURE 14-18. Stays and shrouds.

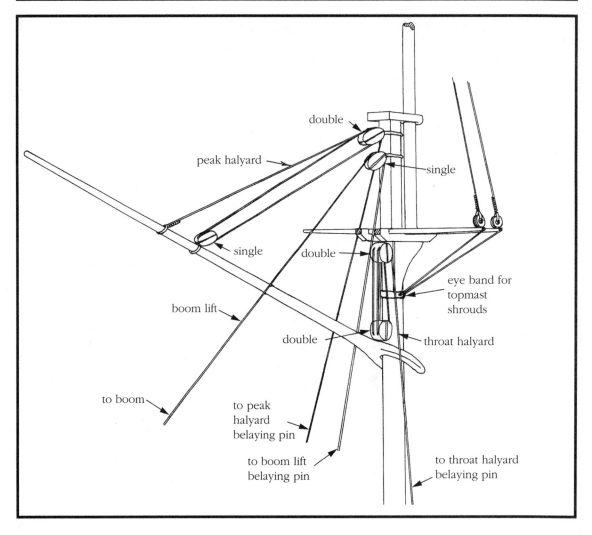

FIGURE 14-19. Rigging of the gaff halyards and the boom lift.

is attached to an eyebolt just behind the eyebolt for the forestay. The lower end of the jib luff secures to a downhaul line that passes through a single block tied to the bottom of the jib stay, and leads back to a belaying pin. Refer again to Figure 14-15.

The sheets are double: one set for a starboard, and the other for a port tack. I suggest that you set these sails to port, making the port sheets taut and the starboard sheets slack. Tackles using single blocks are used to hoist the heads of the headsails.

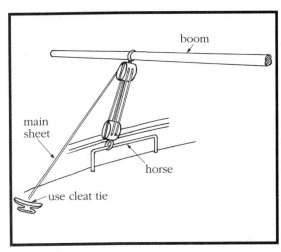

FIGURE 14-20. Main sheet rigging.

FIGURE 14-21. The rigging of a running square sail.

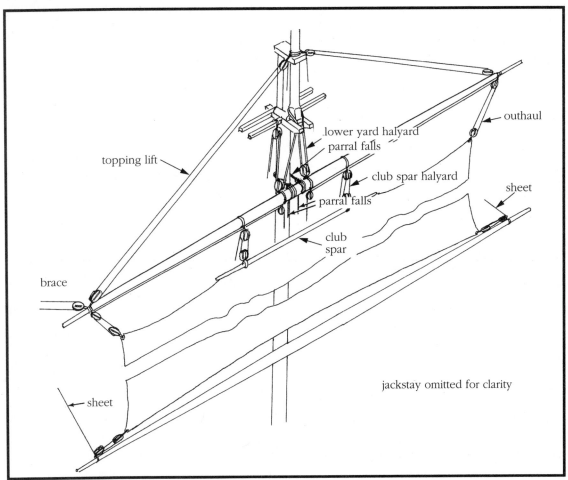

Color scheme and display

Unfortunately, no description of the color scheme for this vessel has survived. She might not have been painted at all, and perhaps you would like to do a natural finish. If you want to paint, however, the following color scheme is based on the conventions of the time:

Hull above the waterline: Dull yellow
Hull below the waterline: Off-white
Wales, channels, and step: Black
Rails, decks, masts, yards, booms, gaff, bowsprit, pumps, skylight, hatch covering, bowsprit bitts, ladders, and catheads: Stained wood
Inside of bulwarks (including quarterdeck bulwark): Dull red
Masthead, including crosstrees and trestletrees: White
Hatch coaming: White

Display the *Lee* either on classic turned brass pedestals or a cradle. A wood-framed case would be appropriate, but be careful to keep the framing small so as not to overwhelm the model.

The *Lee* is an interesting project for any ship modeler. For beginners, it's both interesting and challenging, but still well within their skills.

CHAPTER 15

Postscript

"My wife just came in and took away my manuscript! She said that it was finished and that if I fooled with it any more I'd ruin it!"

— Bewildered author

In closing, let me reiterate a few maxims for the first-time scratchbuilder:

- Don't attempt too much at first. Choose a project within the bounds of your skill, but that requires you to stretch a little.
- Use comprehensive drawings that provide maximum information.
- Don't try to work too fast — no one is imposing deadlines on you.
- If you're totally displeased with some part that you've made, don't hesitate to make a new one. On the other hand, don't expect perfection; just do the best you can.

- If you're not sure of how to do something, check out a reference or ask an experienced fellow modeler. If possible, join a ship-modeling club. Modelers are friendly people; I've never known one who wasn't willing to help.
- Finally, remember that no matter how much you've learned, there's always more. Every time I build a model, I find myself learning something new.

In starting to scratchbuild model ships, you're entering the most challenging and satisfying aspect of ship modeling. Good luck, and smooth sailing!

APPENDIX I

Sources for Plans

The following is a brief list of sources from which you can get plans. In addition, Appendix II lists reference books that might include useful drawings. Another avenue to explore is maritime museums. They often have their own plans, and it's amazing how many museums there are. I've listed a few. For a detailed listing, see *The Naval Institute Guide to Maritime Museums of North America* published by the Naval Institute Press (see Appendix II for information).

Bluejacket Ship Crafters, P.O. Box 425, Stockton Springs, ME 04981
> A small range of plans for their own kits, but also available separately. The drawings are very good. Ask for their catalog.

Catalog of Warship Drawings, Edward Wiswesser
> Small-scale drawings of old and modern naval and merchant vessels. No sailing ships. Some of these drawings require additional research to make a really good model. Available through Taubman's Plans Service (see below).

Dromedary Ship Modeler's Center, 6324 Belton, El Paso, TX 79912
> A good collection of American and foreign plans for all types and periods from a variety of sources.

Floating Drydock, c/o General Delivery, Kresgeville, PA 18333
> Drawings for modern warships, principally American.

Harold M. Hahn, 212 Garden Road, Lyndhurst, OH 44124
> Mr. Hahn has a small range of fine sailing ship drawings he has prepared himself.

Maine Maritime Museum, 243 Washington Street, Bath, ME 04530
> This museum is devoted to Maine's colorful and rich maritime history.

Marine Museum of the Great Lakes, 55 Ontario Street, Kingston, Ontario, Canada K7L 2Y2
> As its name indicates, this museum is dedicated to the shipping of the Great Lakes and St. Lawrence River.

Mariners' Museum, 100 Museum Drive, Newport News, VA 23606
> This museum's collection centers on the Chesapeake Bay area.

Model Engineering and Boat Plans Handbook
> A British catalog containing a wide range of plans for all kinds of ships and boats. Also includes plans for working gasoline, diesel, and steam engines. Available through Taubman's (see below).

Mystic Seaport, Mystic, CT 06335
> A good number of drawings of ships and small craft of particular interest to the Mystic Seaport Museum (New England vessels).

Peabody Museum, East India Square, Salem, MA 01970
> A large collection of maritime documents and drawings. Especially good for New England and the East Coast.

Philadelphia Maritime Museum, 321 Chestnut, Philadelphia, PA 19106
> Specializes in materials relating to the Delaware River and Bay, and to the southern New Jersey area.

Seagull Plans

There aren't many of them, but these plans by Bill Crothers are arguably the best available anywhere. They're both historical documents and works of art. Available through Taubman's (see below).

Smithsonian Institution. There are two sources of plans from the Smithsonian, one for warships and one for other vessels. Write for their catalogs.

The Smithsonian Collection of Warship Plans, Naval Section, Director of Armed Forces History, NMAH 4011/MRC 620, Smithsonian Institution, Washington, D.C. 20560 (catalog costs $6.00)

Ships' Plan List of the Maritime Collection, Ships' Plans, Division of Transportation, NMAH 5010/MRC 627, Smithsonian Institution, Washington, D.C. 20560 (catalog costs $10.00)

Taubman's Plans Service International, 11 College Drive, Box 4G, Jersey City, NJ 07305
> The largest catalog of plans available. A wide range of American and foreign vessels of all types and periods. It's fun just to browse through the catalog.

APPENDIX II

Reference Material

The literature available on ship modeling is enormous, and the references listed below represent only a small fraction of it. They are ones with which I am familiar and feel competent to comment on. By all means, explore the literature on your own. Don't ignore good magazines such as *Model Shipwright, Seaways/Ships in Scale*, and *Model Shipbuilder*. A publication invaluable to any serious ship modeler is the Nautical Research Guild *Journal*. Membership in the NRG brings this outstanding magazine to you regularly.

Albion, R.G. *Naval and Maritime History: An Annotated Bibliography*. Mystic, CT: Marine Historical Association, Inc., 1972.
> A comprehensive listing, with commentary, of the available literature on the subject. Most useful for research purposes.

Bates, A.L. *The Western Rivers Steamboat Cyclopedium*. Leonia, NJ: Hustle Press, 1981.
> If you plan to build a model of an American river steamboat, this is an essential reference. No plans, but many drawings of typical details, and an excellent text.

Boudriot, Jean. *The 74-Gun Ship*. Translated by David H. Roberts. Annapolis, MD: Naval Institute Press, 1986.
> This massive four-volume work is the definitive description of the construction and employment of an eighteenth-century ship-of-the-line. It's expensive, but as a reference work it's unsurpassed. Extensively illustrated with marvelous plans and detailed drawings showing every detail of the ship and its equipment.

Chapelle, H.I. *American Sailing Craft*. Camden, ME: International Marine Publishing Co., 1975.
> A review of typically American sailing craft extant in the 1920s and '30s. Includes many excellent drawings.

———. *American Sailing Ships*. New York: W. W. Norton, 1935.
> An extremely useful history of the design of American sailing ships by America's greatest marine historian. Contains many useful drawings from which models can be made.

———. *Boatbuilding*. New York: W. W. Norton, 1941.
> If you're planning to build a model of a small boat, this book is essential. It tells you in clear terms how the real thing is built, and is well illustrated. The chapter on lofting alone is worth the price of the book.

———. *The History of the American Sailing Navy*. New York: W. W. Norton, 1949.
> A comprehensive history of the design of American sailing warships. Contains a wealth of background information and many authoritative drawings suitable for modeling.

———. *The National Watercraft Collection*. Camden, ME: International Marine Co., 1976.

Contains descriptions of selected models in the Smithsonian Institution with information about the original ships. Includes a number of drawings and many photos of use to the modeler.

————. *The Search for Speed Under Sail, 1700 - 1855*. New York: Bonanza, 1967.
A review of the history and design of American and pre-Revolutionary British ships built for speed. Contains many excellent drawings from which models can be built.

Chapman, F.H. *Architectura Navalis Mercatoria*, London: Adland Coles Ltd.
This is a modern facsimile reprint of an eighteenth-century Swedish book by the greatest naval architect of the period. It contains hundreds of exquisite drawings of eighteenth-century ships and boats of all types. Many of the drawings are suitable for model work, but there are no rigging plans.

Coker, P.C., III. *Building Warship Models*. Charleston, SC: Cokercraft, 1974.
Although there are no drawings in this book from which models can be made, it contains a wealth of information on how to construct the elements of a modern warship. Very well illustrated with drawings and photos.

Davis, C.G. *The Ship Model Builder's Assistant*. Largo, FL: Edward W. Sweatman, 1970.
This is a reprint of a book originally published in the 1920s. It's considered a classic in the field, and contains much useful information about how to model ships, as well as tables of dimensions and proportions. Especially good for rigging. Well illustrated, but there are no drawings from which models can be made. Some of the recommendations and procedures are out of date.

————. *The Built-up Ship Model*. New York: Dover, 1989.
This is a reprint of a book originally published in 1933. It describes in detail the building of a model of the Revolutionary War brig *Lexington*. No rigging work is described (this was an unrigged model), but spar dimensions are given. Extensively illustrated, and an excellent reference.

————. *Ship Models and How to Build Them*. New York: Dover, 1986.
This is a reprint of a book originally published in 1925. It describes the building of a lift model. Much of the material is outdated, but it remains an interesting reference. Includes a foldout lines drawing of the clipper *Sea Witch*.

Falconer, W. *Falconer's Marine Dictionary*. New York: Augustus M. Kelley, 1970.
A reprint of an eighteenth-century book that defines the nautical terminology of the period. A tremendous help in understanding the ships of that time. No plans, but many excellent drawings.

Grimwood, V.R. *American Ship Models*. New York: Bonanza, 1942.
Tells how to build 12 models of American sailing ships using the lift method. Contains good drawings that you can modify for plank-on-frame construction. This is quite a useful book if you can find it.

Harland, J. *Seamanship in the Age of Sail*. Annapolis: Naval Institute Press, 1987.
A splendid book, extremely well illustrated, on how ships were actually sailed. No plans, but a great deal of information on how rigging was set up and used.

Hough, R. *Fighting Ships*. London: Michael Joseph, 1969.
A well-illustrated history of warships from earliest to modern times. No plans, but much background information.

Howard, F. *Sailing Ships of War, 1400-1960*. New York: Mayflower, 1979.
An history of sailing warships with many illustrations and invaluable tables of data. Written with the modeler in mind. No plans, but useful for research purposes.

Landström, B. *The Ship*. Garden City: Doubleday, 1961.
Perhaps the most beautifully and authorita-

tively illustrated book in existence on the history of ships. Progresses from prehistory to modern times. Although not oriented toward the modeler's needs, it contains much useful information, and should be in the library of everyone who loves ships.

Lavery, B. *The 74-Gun Ship* Bellona. Annapolis: Naval Institute Press, 1985.
A detailed analysis of the design and construction of an eighteenth-century ship-of-the-line. If you can't afford the Boudriot book, this is the next best thing. Well illustrated with detailed drawings. This is a valuable reference work, since much of it can be applied to other ships of the period. This is one book in a series titled *The Anatomy of the Ship*. Each book in the series is written by a different author and treats a different ship in detail. They are all highly recommended.

Lyon, H. *The Encyclopedia of the World's Warships*. New York: Crescent, 1978.
Brief histories, statistics, and small three-view drawings of modern warships from World War I to the 1970s. Very interesting, and useful as a general reference.

Mastini, F. *Ship Modeling Simplified*. Camden, ME: McGraw Hill, 1990.
This book is written for the novice who wants to build kits. Well written and illustrated. Includes a dictionary of Italian nautical terms, which is useful to the kit builder because so many kits are Italian made.

McKee, E., *Clenched Lap or Clinker*. Greenwich: National Maritime Museum.
This book, really not much more than a pamphlet, is the most comprehensive and valuable work I've seen on the construction of lapstrake boats. Well illustrated.

McNarry, D. *Shipbuilding in Miniature*. London: Percival Marshall, 1955.
This book by the master of miniature modeling is fascinating, even if you don't plan to build a miniature. The techniques are useful for making small fittings on models of any scale.

Mondfeld, W. *Historic Ship Models*. New York: Sterling Publishing Co. Inc., 1985.
A handsome book, translated from the German and originally published in 1977. Good for general reference. The many illustrations show various styles of fittings. The book includes reference tables as well as a list of nautical terms in several languages.

Nautical Research Guild. *Ship Modeler's Shop Notes*. Bethesda: Nautical Research Guild, 1979.
A collection of short articles on various aspects of ship modeling. This is a "must have" for the serious ship modeler, and offers valuable assistance to the beginner.

Preston, Lyon, and Batchelder. *Navies of the American Revolution*. Englewood Cliffs, NJ: Prentice Hall, 1975.
A well-illustrated book about the personnel, ships, and equipment of the American Revolution. Of limited value to the modeler, it is nevertheless recommended for its evocation of the naval environment of the period.

Roth, M. *Ship Modeling from Stem to Stern*. Blue Ridge Summit, PA: TAB, 1988.
A well-illustrated and up-to-date book. Excellent reference for general techniques. Contains tables of proportions.

Schairbaum, A.W. *A Marine Glossary for the Ship Modeler*. St. Charles, IL, 1990.
Published by the author, this is a useful small dictionary of nautical terms. Comprehensive illustrations are included.

Smith, R. H. *The Naval Institute Guide to Maritime Museums of North America*. Annapolis, MD: Naval Institute Press, 1990.
Lists approximately 300 museums in the United States and Canada, with descriptions of hours, fees, services, and other information.

Takakjian, P. *32-Gun Frigate* Essex. Cedarburg, WI: Phoenix, 1985.
This large pamphlet describes the construction of a model of the *Essex*. The model can be built from this book; full-size plans must be purchased from model outlets such as

Dromedary. The book is also useful as a reference to construction techniques.

Underhill, H.A. *Plank-on-Frame Models.* Glasgow: Brown, Son, and Ferguson, 1971.

This two-volume set centers on the construction of a model of the nineteenth-century brigantine *Leon,* but it's much more than a one-ship book. It's by far the best exposition of plank-on-frame techniques, and is also extremely valuable as a reference for masting and rigging. Very well illustrated with clear line drawings. If you own no other "how to" book, you should own this one.

APPENDIX III

Tools and Materials Sources

The following is a short list of sources for tools and materials. You should be able to find whatever you need from these sources, but don't overlook your local hobby shop.

Anchor Tool and Supply Company, Inc., 231 Main St. (Rte 24), Chatham, NJ 07928
    Every possible kind of useful hand tool. Highly recommended.

Bluejacket Ship Crafters, P.O. Box 425, Stockton Springs, ME 04981
    A wide range of fine fittings. They also carry the more common modeler's hand tools, some woods, and brass shapes.

Dockyard Model Company, P.O. Box 80656, Colorado Springs, CO 80933
    Tools and fittings.

Dromedary Ship Modeler's Center, 6324 Belton, El Paso, TX 79912
    Tools and fittings.

Hobby Products Company, 2757 Scioto Parkway, Columbus, OH 43026-2334
    The best source for the Unimat lathe and accessories.

The Lumberyard, 6908 Stadium Drive, Brecksville, OH 44141
    A line of hard-to-find woods, including boxwood and pear.

Micro-Mark, 340 Snyder Ave., Berkley Heights, NJ 07922
    Small hand and machine tools. Ask for their extensive catalog—you'll find it almost indispensable.

U.S. General, 100 Commercial St., Plainview, NY 11803
    Mail-order catalog, and stores in many cities. A large inventory of all kinds of hand and power tools. Oriented toward the general tool market, but you'll find many items useful in modeling.

Wood Carvers' Supply, Inc., P.O. Box 8928, Norfolk, VA 23503
    Everything for the woodcarver. Many tools appropriate for model building. You should definitely have their catalog.

APPENDIX IV

Glossary

This glossary includes definitions for all terms used in this book that might be unfamiliar to some readers. In addition, it includes other common terms that you might encounter in other sources.

airport — A round glazed window that can be opened for ventilation. Shown in Figure 11-1. Often referred to as a *porthole*.

backstay — An item of standing rigging that leads from a channel or a fastening on the deck to the masthead of the topmast or higher. A running backstay is equipped with block and tackle so that it can be easily tightened or slacked off.

bark — A ship with at least three masts, all of which are square rigged except the aft mast, which is fore-and-aft rigged.

barkentine — A ship with at least three masts, of which the fore mast is square rigged and all others are fore-and-aft rigged.

base line — A line on a drawing that's used as a basic reference for the construction of other lines. Shown in Figure 2-6. See also *reference lines*.

beam shelves — See *deck beam*.

bearding line — A line that indicates the intersection of the planking with the deadwood, thus showing where the deadwood must be cut away to receive the ends of the planks. Shown in Figure 2-8.

bees — Pieces of wood placed on the bowsprit just behind the bowsprit cap. The bees have holes through which the forestays are passed, preventing the stays from shifting. Shown in Figure 9-3.

belaying pin — A fitting that resembles a policeman's billy club. Belaying pins fit into holes in the pinrails and are used to secure the free ends of running rigging. The ends of the line can be quickly freed by pulling the belaying pins out of the rails. Shown in Figure 9-23.

bent frames — Relatively thin strips of wood bent to the inside shape of the hull and attached after the hull planking has been built up over a mold. Used on small boats for which the frames are not built up. See also *frame*.

bilge keel — On steel ships, false keels placed along the sides of the ship at the turn of the bilge that assist in stabilizing the ship. Shown in Figure 7-9.

bilge stringers — Longitudinal reinforcing planks located along the insides of the frames at the turn of the bilge. Shown in Figure 6-7.

billboard — A heavy piece of wood placed so as to protect the sides of the ship from being chafed by the stowed anchor. Also referred to as *noble wood*.

bitts — A pair of vertical posts, made of wood or metal, used to secure the end of a cable.

boat — A small craft capable of being carried on board a larger vessel.

bobstay — An item of rigging running from the end of the bowsprit to the cutwater. Frequently made of chain. Shown in Figure 14-15.

body plan — The portion of a drawing that portrays the lines of the ship as seen from fore and aft. Shown in Figure 2-1. Also called the *section view.*

bolster — A quarter-round piece of wood placed on top of a trestletree over which the shrouds pass. Shown in Figure 9-5.

bolt rope — A rope attached to the edge of a sail. The bolt rope strengthens the sail and provides a strong attachment for lashings or hoops that fasten the sail to the mast or yard. Shown in Figure 10-4.

bomb ketch — A ketch-rigged warship designed to carry heavy sea mortars used in shore bombardment. The mortar was emplaced forward of the mainmast, and the forestays of the vessels were of heavy chain to withstand the muzzle blast of the guns. Used in the eighteenth and early nineteenth centuries.

boom — A spar that supports the foot of a fore-and-aft sail. Shown in Figure 9-7. See also *jib boom* and *studdingsail boom.*

boom iron — A hardware fitting attached to the end of a yard that supports the studdingsail boom. Shown in Figures 9-6 and 9-19.

bow — The forward end of the ship.

bowsprit — A spar protruding from the bow of the ship that supports the forestays and the tackle for handling the headsails. Shown in Figure 9-3.

brace — A tackle attached to the ends of the yards. Used to adjust the angle of the yards according to the wind and course direction.

breast hooks — Horizontally placed knees across the inside of the stem of a ship that reinforce the bow structure. Shown in Figure 6-8.

brig — A two-masted vessel; both masts are square rigged.

brigantine — A two-masted vessel in which the foremast is square rigged and mainmast is fore-and-aft rigged.

bulkhead — As used in this book, a structure at right angles to the centerline of the ship that has the shape of a section. Shown in Figures 6-13 and 6-15.

bullseye — A round rigging fitting with a single hole in the center and a groove around the edge. Used to provide a non-wearing connection between two lines. Shown in Figure 9-14.

bulwark — A portion of the hull planking above the uppermost deck.

buttock line — The shape of a vertical plane passed through the hull parallel to the vertical center plane of the hull. On a drawing, the shapes of the buttock lines appear on the elevation view, but they appear as straight lines on the body plan and plan view. Shown in Figure 2-3.

camber — The transverse curvature of the deck.

cant frame — A half-frame located near the bow or stern set at an acute angle to the centerline. Cant frames reduce the amount of bevel required on a frame where the curvature of the planking becomes extreme near the bow and stern. They also provide greater structural strength in these areas. Shown in Figure 2-17. See also *frame.*

cap — A fitting used at a masthead or at the head of the bowsprit to link the mast to the mast above or, in the case of the bowsprit, to link the bowsprit to the jib boom. Shown in Figures 9-3, 9-4, and 9-5.

cap rail — A rail that "caps" the top of the timberheads and planking.

caravel — A small sailing craft, usually lateen rigged, of the fifteenth century. The *Nina* and *Pinta* were caravels, rigged with square sails for the Atlantic voyage.

carlins — Longitudinal timbers between deck beams that form the sides of deck openings. Shown in Figure 6-8.

carrack — A name applied to large northern European ships of the 14th and 15th centuries. They were lapstrake built and distinguished by an extremely heavy mainmast.

carvel-planking — A method of planking a ship in which the planks are laid edge to edge, as opposed to lapstrake planking, in which the plank edges overlap.

cat — An eighteenth-century ship distinguished by a narrow stern, projecting quarters, and a blunt bow with no beakhead or stem ornamentation.

catboat — A small, shallow draft boat, usually equipped with a centerboard. It has one mast placed right in the bows, and is usually gaff rigged. Very popular in the 19th and early 20th centuries as pleasure and racing craft.

cathead — A heavy timber placed near the bow of a ship that's used as a crane for handling the anchor. Shown in Figure 14-11.

centerline — A line used to divide the plan or section view of a drawing into two equal parts.

chain plate — A hardware fitting used to fasten the stroppings of the lower deadeyes to the hull.

channel — Broad planks fixed to the sides of a ship to position the lower deadeyes and chain plates. On warships there was a single channel on each side for each set of shrouds; on nineteenth-century merchantmen, however, there were usually two on each side, one above the other.

cheeks — See *hounds*.

clench-nail — A type of lapped hull construction. See also *lapstrake*.

clinker-built — A type of lapped hull construction. See also *lapstrake*.

clove hitch — A knot used to attach ratlines to the shrouds. Shown in Figure 9-21.

cog — A type of merchant ship used in the 13th and 14th centuries by the Hanseatic League. They were straight ended, of lapstrake construction, and had a deep draft. They carried one square-rigged mast.

counter — The part of a ship's stern under the transom.

course — The term applied to the principal sails of a ship; for example, the lowest square sail on the main mast is called the main course.

cowhorn — A small double-ended, lapstrake-built boat with two gaff-rigged masts. The best of these were to be found on Block Island in the nineteenth century. They were extremely seaworthy.

crosstree — Timbers placed at the bottom of a masthead at right angles to the centerline that spread the shrouds of the mast above and support the planking of a top platform. Shown in Figure 9-5.

cutwater — The foremost timbers of the stem assembly. These timbers are the first part of the ship to cut the water, hence the name.

davits — Small cranes usually used to hoist ship's boats, but sometimes used as anchor hoists. Shown in Figure 13-14.

deadeye — A round rigging fitting, usually with three holes and a groove around the rim. Some ships prior to the seventeenth century had triangular deadeyes. They're used in pairs at the lower ends of the shrouds to make it possible to adjust the tension of the shrouds. When wire rope rigging came into use, they were replaced with turnbuckles. Shown in Figure 9-26.

deadrise — The angle of rise of a ship's bottom from the keel to the turn of the bilge.

deadwood — Timbers placed at the angles between the keel and stem and sternpost that strengthen the structure. The aft deadwood also provides a landing place for the aft ends of the planking. Shown in Figure 2-9.

deck beam — Timbers placed at right angles to the centerline that support the deck planking. These timbers usually have a slight curve (camber), which gives the deck a curved, water-shedding surface. The ends of the deck beams rest on timbers called *beam shelves* that are attached to the inner surfaces of the frames. Shown in Figures 6-7 and 6-8.

diagonal — A line on a drawing that describes the shape of the hull along a plane set at an angle to the vertical plane of the centerline. On the drawing, diagonals are seen as slanted straight lines on the body plan. Their true shapes are usually plotted on their own baseline and superimposed over the plan view. Shown in Figures 2-4 and 2-13.

elevation — On a drawing, the view that shows the ship as seen from its side. Shown in Figure 2-1. Also called the *profile*.

eye band — A metal band with one or more eyes protruding from its sides. Used on spars to provide attachment points for rigging. Shown in Figure 9-17.

eyebolt — A hardware fitting consisting of a metal rod bent into an eye at one end. One of the most ubiquitous items on a sailing ship, they are used to secure the standing (that is, not moving) ends of lines. Often eyebolts are fitted with a ring through the eye that provides greater flexibility at the end of the line. In this case they are called *ringbolts*. Shown in Figure 9-15.

fid — A short piece of wood or metal passed through the heel of a top or topgallant mast that prevents the mast from slipping downward. Shown in Figure 9-5.

fluyt — An eighteenth-century Dutch ship with a round stern, narrowing above.

fore-and-aft sail — A sail with its forward edge, or luff, attached to a mast. The foot of the sail is secured by a boom, whose forward end is hinged to the mast by jaws or by a gooseneck fitting.

frame — An assembly that has the shape of a section. Frames, which are attached to the keel, provide the framework to which the planking is attached. Full and half-frames are placed at right angles to the centerline. Cant frames are placed at acute angles to the centerline. Sometimes called *ribs*. See also *bent frame*.

frigate — A three-masted sailing warship normally rated between 28 and 44 guns. Also, as late as the mid-eighteenth century, a type of hull construction.

futtock — A component of the frame assembly. Frames are not sawn in one piece, but are built up in two layers. Several futtocks make up each layer, and the layers are made up so that the joints in each layer are overlapped by the futtocks of the other layer.

gaff — A spar that supports the head of a trapezoidal fore-and-aft sail. Shown in Figure 9-7.

gaff sail — A trapezoidal fore-aft-sail. Shown in Figure 10-1.

galleass — A large galley distinguished by a heavily armed turret-like structure in the bow.

galleon — A Spanish term used in the 16th and 17th centuries to describe a three-or four-decked warship. Later used as a generic term to describe a large, usually four-decked, merchantman.

galley — A ship designed to be propelled by oars. Most galleys were also capable of carrying considerable sail, and by the eighteenth century the term was applied to any sailing ship that could use oars as auxiliary power.

gammoning — The lashing that holds the bowsprit down to the top of the stem.

garboard strake — The strake that fits into the keel rabbet. Shown in Figure 6-7. See also *strake*.

gooseneck — A hardware fitting that supports the forward end of a boom. Shown in Figure 9-18.

gore line — A line that represents the intersection of two patterns of coppering. The plates must be cut into triangular or trapezoidal shapes at this line. See Figures 7-2 and 7-3.

gudgeon — The female portion of a rudder hinge. The male component is called a *pintle*. Shown in Figure 7-10.

gunwale — The topmost plank of the hull. Traditionally pronounced "gunnel."

guy — An item of standing rigging used to support a spar. Examples are the martingale guy and the jib boom guy.

hag boat — An eighteenth-century northern European merchantman whose planking was carried up to a beam just under the taffrail, with no counter.

half-breadth plan — See *plan view.*

half-frame — A frame that, because of its location (such as adjacent to the deadwood), can't be mounted directly on the keel. In such a case, port and starboard halves must be made and mortised into the deadwood or the stem assembly.

hanks (jib) — Small rings used to attach the luff of a headsail to its stay.

hawse hole — An opening in the bow of a ship through which an anchor cable passes.

head sail — A triangular sail located forward of the foremast. These sails are attached to stays that lead from the foremast to the bowsprit, jib boom, or flying jib boom. They are classified, according to their positions, as forestaysail, jib sail, or flying jib sail.

heart — Similar to a bullseye, except that a heart is triangular in shape. Shown in Figure 9-14.

horse — A low rail to which the sheets of a fore-and-aft sail are attached. In modern terminology, this is called a *traveller.*

hounds — Brackets mounted on each side of a mast just below the masthead that support the trestletrees. Shown in Figures 9-2 and 9-5. Also called *cheeks.*

jackstay — A metal rod attached to the top of a yard by eyebolts to which the square sails are bent. Shown in Figure 9-6. Also, in some types of rig, a line running down the front of a mast to which the lower yard is attached. Shown in Figure 14-17.

jaws — Assemblies at the forward end of a boom or gaff. The jaws form a two-pronged fork that fits around the mast and is secured by a parral. Shown in Figure 9-7.

jib boom — A spar that extends the bowsprit. Flying jib booms extend the jib boom. Shown in Figure 9-3.

jib sail — See *head sail.*

jigger mast — The fourth mast of a ship.

joggling — See *nibbing.*

keel — The bottommost timber of the ship. It can be considered the ship's backbone, upon which the stem, sternpost, and frames are mounted.

keelson — A timber mounted on top of the frames above the keel. It locks the frames in place and, with the keel, forms a strong girder. Shown in Figure 6-1.

ketch — A two-masted vessel in which the aftermast is notably shorter than the foremast. Sometimes referred to as ship without a foremast.

kevil — An early type of V-shaped cleat located on the inside of the bulwarks. Used for securing the main or fore sheets.

knees, hanging — Vertically oriented knees that support the ends of the deck beams. Shown in Figure 6-7.

knees, lodging — Horizontally oriented knees that brace the ends of the deck beams against fore and aft movement. Shown in Figure 6-8.

lanyard — The line used to join the upper and lower deadeyes. Shown in Figure 9-26.

lapstrake — A type of hull construction in which the planks, or strakes, slightly overlap each other. Lapstrake hulls are built over molds, and bent frames are inserted after the hull planking is complete. Shown in Figures 6-20 and 6-21.

leech — On a gaff or headsail, the after edge of the sail. On a square sail, the side edges of the sail. Shown in Figure 10-1.

lift model — A model in which the hull is built up with layers (lifts) of wood shaped like waterlines and then carved to the final shape.

lights — Traditionally, the windows in a ship.

limber strakes — Longitudinal planks along the

insides of the frames near the keel. Shown in Figure 6-7.

load waterline — The waterline that lies at the surface of the water when the ship is loaded. Often abbreviated *l.w.l.*

lofting — The process of translating a drawing into full-size patterns for making the components of the hull.

luff — On a gaff or headsail, the forward edge of the sail. This edge is attached to the mast or, in the case of headsails, to the stay. Shown in Figure 10-1.

margin plank — The outermost deck plank. It conforms to the curve of the hull. Shown in Figures 6-18 and 6-19.

martingale — A spar extending downward from the end of the bowsprit that supports the guys, which in turn hold down the jib boom. It can be either single or in the shape of an inverted V. Shown in Figure 9-3. Also known as the *dolphin striker.*

masthead — The topmost part of a mast, from the bottom of the trestletrees to the cap. The masthead overlaps the foot of the mast above it. Shown in Figure 9-5.

mast hoop — A ring that loosely surrounds a mast. Used as an attachment point for the luff of a fore-and-aft sail. Shown in Figure 9-20.

midsection — The widest section of a ship.

mold — A structure over which small boats are planked prior to the installation of the bent frames. Shown in Figures 6-22 and 6-24.

nao — A Spanish term used from the 14th to 16th centuries to denote a large ship. Columbus referred to the *Santa Maria* as a nao.

nibbing — The process of fitting the ends of deck planks into the margin plank. Shown in Figure 6-19.

parral — An assembly of rope and large wooden beads used to secure a yard to a mast, or to secure the jaws of a boom or gaff to a mast. In some cases rope only, with no beads, is used. Shown in Figures 9-6 and 9-7.

partners (mast) — The heavy timbers incorporated into the deck framing that support the mast where it passes through the deck. Shown in Figure 6-8.

peak halyard — The tackle used to adjust the set of a gaff. It usually consists of two or more blocks attached to the gaff, and one or more blocks secured to the aft side of the masthead. The run of the tackle varies, and double and/or single blocks can be used. See Figure 14-17.

perpendiculars — Reference lines placed at right angles to the base line at the bow and stern. They're usually placed to intersect the foremost point of the stem rabbet and the aftmost point of the sternpost rabbet. Shown in Figure 2-6.

pink — A eighteenth-century ship with a very narrow stern.

pinky — A two-masted fishing schooner of the nineteenth century common to New England and Nova Scotia. It was characterized by a double-ended hull and a very narrow transom.

pinnace — A small vessel equipped with both oars and sails. Usually two masted and schooner rigged. Also, an eight-oared boat.

pin rail — A rail placed inside the bulwarks and drilled with holes to receive belaying pins. A pin rail placed at the foot of a mast is called a *fife rail.* Shown in Figure 14-13.

pintle — The male portion of a rudder hinge. The female component is called a *gudgeon.* Shown in Figure 7-10.

plan view — On a drawing, the view that shows the lines as seen from the top. Ordinarily, only half the view is shown; this is called a *half-breadth plan.* Shown in Figure 2-1.

plank-on-bulkhead — A model in which the hull is built up with solid bulkheads, rather than with realistic frames, and then planked.

plank-on-frame — A model in which the hull is built up with realistically made frames, and then planked as a real ship would be.

plank sheer — The line on a drawing that describes the run of the top edge of the planking.

profile — See *elevation.*

rabbet — A groove cut into the keel, stem, and sternpost that receives the planking. Shown in Figure 2-7.

ratline — Ropes lashed horizontally to the shrouds that serve as ladder steps.

reef band — A heavy strip of cloth sewn to a sail that reinforces the sail where the reef lines are attached. Shown in Figures 10-1 and 10-5.

reef line — Short lengths of rope attached to a sail that allow the sail area to be reduced. This is accomplished by lashing them under the boom for a gaff sail, or under the foot of the sail for other types. Shown in Figure 10-5.

reference lines — The lines on a drawing from which measurements are made. The baseline, centerlines, and perpendiculars are all considered reference lines. Shown in Figure 2-6.

ribs — See *frame.*

ringbolt — See *eyebolt.*

room and space — The distance between the centers of two frames; or, in other words, the space taken up by one frame plus the distance to the next frame. Shown in Figure 2-10.

running rigging — The lines on a ship that run through blocks or are otherwise movable.

scarf joint — A joint for joining the ends of timbers. Typically shaped like an elongated N. Shown in Figure 6-1.

schooner — A vessel with at least two masts, all of which are fore-and-aft rigged. Three-masted schooners are called *tern schooners.* Three-, four-, and five-masted schooners were common. A few six-masters were built, and one—the *Thomas Lawson*—was a seven-master. During the 18th and early 19th centuries, two-masted schooners fitted with square sails on the foremast were called *topsail schooners.*

scupper — An opening at the intersection of a bulwark and a deck that allows water on the deck to drain.

section — The shape of a vertical plane passed through the hull at right angles to the centerline. Their true shapes are seen on the body plan, but they appear as straight lines on the elevation and plan views. Shown in Figure 2-5.

seizing — A method of lashing the end of a line, or of lashing two lines together. Shown in Figure 9-25.

sheer — In the elevation view of a drawing, the curve described by the top of the planking.

sheer plank (sheer strake) — The plank that follows the line of the sheer just below the main deck.

ship — Strictly speaking, a sailing vessel that has at least three masts, all of which are square rigged.

ship-of-the-line — The largest sailing warships, which were considered fit to lie in the line of battle. Sixty-four guns was considered the minimum.

shroud — An item of standing rigging that supports a mast. There are also bowsprit shrouds.

sloop — In modern terms, a vessel with one mast, fore-and-aft rigged. In reference to sailing warships, a ship smaller than a frigate. Sloops-of-war could be ship rigged.

snow — A brig in which the luff of the trysail was set on a small mast (the snow mast) doubled to the after side of the main mast.

spanker — The gaff-rigged sail on the mizzenmast of a ship.

spar — A generic term that includes masts, yards, bowsprit, booms, and gaffs.

spritsail yard — A yard that crosses the bowsprit. Until the eighteenth century, this yard supported a sail, but when jib sails were introduced, this became unnecessary. Eventually the sail was abandoned, and the yard was used only to support the jib boom guys. Shown in Figure 9-3.

square sail — A sail, often trapezoidal, that is attached at the head to a yard. Shown in Figure 10-1.

standing rigging — The fixed rigging that supports the masts and the bowsprit.

station — On the elevation and plan views of a drawing, the lines that indicate where the sections shown in the body plan are located.

stay — Generally, any item of standing rigging other than shrouds. Some stays (such as for the jib boom) are referred to as *guys*.

stealer — A short plank used to adjust the run of the planking where the planking tends to spread apart (usually near the stern) or to come together too closely (usually near the bow). Shown in Figures 6-10 and 6-11.

stem — The forward vertical extension of the keel. Shown in Figure 6-1.

stern — The after part of the ship.

sternpost — A vertical, or nearly vertical, timber rising from the after end of the keel. Shown in Figure 6-1.

strake — A single line of planking from stem to stern. Except in small boats, a strake is made up of several planks.

studdingsail — A long, narrow sail used to extend the area of a sail. These sails give clipper ships the appearance of having wings.

studdingsail boom — A spar installed at the end of a yard to support a studdingsail. The spar may be extended or retracted. Such spars can also be found on the booms of clipper schooners. Shown in Figure 9-6.

tabling — The hem of a sail. Also, reinforcing layers of cloth applied to the lower part of the leech of a fore-and-aft sail. Shown in Figures 10-1 and 10-3.

taffrail — The rail across the top of the transom.

template — A pattern used to check the shape of a hull. Shown in Figure 6-28.

throat halyard — The tackle that raises a gaff. Usually consists of a double block just behind the jaws of the gaff and another secured to the center of the aft crosstree. Shown in Figure 14-19.

thumbcleat — A small piece of wood placed on a spar to keep a line lashed around the spar in place. Shown in Figure 9-7.

thwart — On a small boat, an interior plank leading from one side of the boat to the other. Thwarts both strengthen the boat and serve as seats. Shown in Figure 6-27.

timberhead — The topmost part of a frame that extends above the top deck.

transom — The planking across the stern of a ship or boat. Transoms vary from a simple single plank on some small boats to very large, richly ornamented structures with windows.

traveller — See *horse*.

treenail — A wooden peg used in lieu of a nail to fasten a plank. Usually pronounced "trunnel."

trestletree — Timbers placed parallel to the centerline on each side of a mast, supported by the hounds. The trestletrees, in turn, support the crosstrees. Shown in Figure 9-5.

trysail — On a brig, the gaff sail on the mainmast.

tumblehome — The inward slope of the topsides of a ship.

vang — The tackle that runs from the outer end of the gaff to the deck to control the set of the gaff sail. Vangs are always in pairs, port and starboard.

wale — An extra thick band of planking along the sides of a ship that strengthens the structure. On older ships (early nineteenth century and before), the wales were visible, forming wide ridges along the hull. Later they were planed off so that, although present, they could not be seen.

waterline — The shape of a horizontal plane passed through the hull at right angles to the vertical center plane of the hull. They appear as straight lines on the elevation view and body plan, and their true shapes appear on the plan view. Shown in Figures 2-2 and 2-19.

waterline model — A model in which the underbody is omitted and the hull is built only from the waterline up.

waterway — A timber placed at the intersection of the deck planking and the bulwark. On merchantmen, this timber appeared as a raised margin plank; on warships, its exposed face slanted from the deck to the bulwark. Shown in Figure 6-18.

whisker boom — A spar mounted laterally on the bowsprit used to spread the jib boom guys. It was the successor to the spritsail yard.

xebec (or chebeck) — A fast sailing lateen-rigged vessel originating with the Barbary pirates. It was derived from the earlier Mediterranean galleys, but did not use oars for propulsion.

yard — A spar that crosses a mast or bowsprit to support a square sail.

yawl — A two-masted vessel in which the after mast is very short in comparison to the main mast.

yoke — A horseshoe-shaped item used to secure forestays to the bowsprit. Shown in Figure 9-14. Also, a wooden fitting attached to the back of a yard where it rests against the mast. Shown in Figure 9-6.

APPENDIX V

Working Drawings:

Brockley Combe *and* Lee

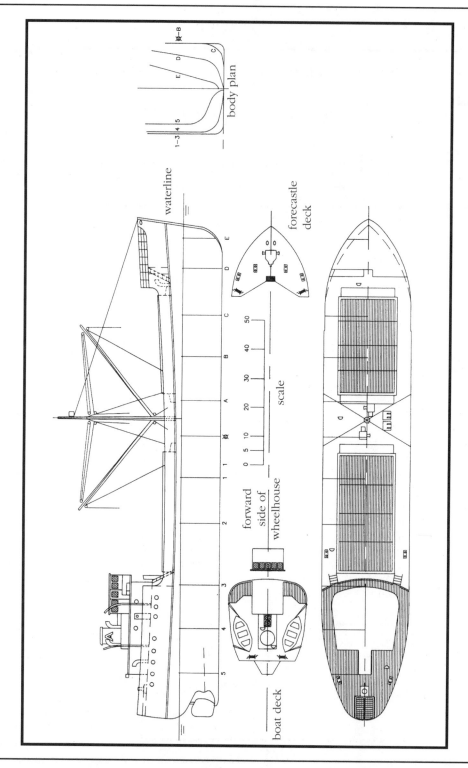

body plan

waterline

forecastle
deck

forward
side of
wheelhouse

scale

boat deck

A working drawing for the *Brockley Combe.*

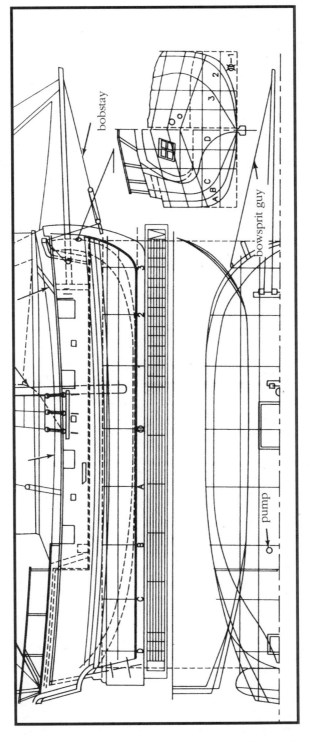

Working drawing for the *Lee* based on a drawing from *The History of the American Sailing Navy*, W. W. Norton & Co.

Index